# *Natural Regulation of Animal Populations*

# *Natural Regulation of Animal Populations*

EDITED BY

## *Ian A. McLaren*

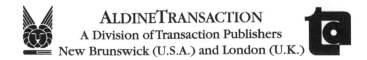

**ALDINETRANSACTION**
A Division of Transaction Publishers
New Brunswick (U.S.A.) and London (U.K.)

Second paperback printing 2009
Copyright © 1971 by Transaction Publishers, New Brunswick, New Jersey.

All rights reserved under International and Pan-American Copyright Conventions. No part of this book may be reproduced or transmitted in any form or by any means, electronic or mechanical, including photocopy, recording, or any information storage and retrieval system, without prior permission in writing from the publisher. All inquiries should be addressed to AldineTransaction, A Division of Transaction Publishers, Rutgers—The State University, 35 Berrue Circle, Piscataway, New Jersey 08854-8042. www.transactionpub.com

This book is printed on acid-free paper that meets the American National Standard for Permanence of Paper for Printed Library Materials.

Library of Congress Catalog Number: 2006044699
ISBN: 0-202-30876-6
Printed in the United States of America

Library of Congress Cataloging-in-Publication Data

Natural regulation of animal populations / [edited by] Ian A. McLaren.
    p.  cm.—(Social problems and social issues)
  Originally published: 1st ed. New York : Atherton Press, 1971.
  Includes bibliographical references and index.
  ISBN 0-202-30876-6 (alk. paper)
  1. Animal populations.  I.  McLaren, I. A. (Ian Alexander), 1931-

QL752.N38   2007
591.7'88—dc22                                                    2006044699

# Contents

# Introduction

IAN A. McLAREN

SOME GENERAL CONSIDERATIONS OF POPULATION GROWTH,
AND THE SCOPE OF THIS BOOK

Lotka (1925), in his classical book on mathematical biology, epitomized the growth of a single species population in his "fundamental theorem of the kinetics of evolution":

$$\frac{dN_1}{dt} = f(N_1, N_2, N_3, \ldots; \quad P_1, P_2, P_3 \ldots). \quad (1)$$

Here $N_1$ is the quantity of interest (size of a population of a species) which changes as a function of the other Ns as influential quantities (other species, nutrient concentrations, and so on) and the Ps representing physical, chemical, and topographical parameters of the environment. Clearly this equation, although not altogether frivolous, is not very useful, since it defines an unlimited range of occupations for people who call themselves ecologists. In addition to vast numbers of quantities and

1

parameters, the problems facing ecologists are compounded by the fact that the various species (Ns) in an environment are themselves interrelated by functions like that of equation 1, and by the possibility of evolutionary changes of the Ns and Ps involved. The constraint noted by Lotka, that $\Sigma N_i$ as matter or energy in a closed system can be taken as constant, is of little help. Nevertheless, it is clear that species cannot be studied in isolation, and some of the most active and original workers in modern ecology are looking into the organization of large systems. Fortunately, there are certain regularities in such systems. One approach is through food-chain and food-web studies, which are extensions of prey-predator models and have been treated as such since the pioneer work of Volterra (see D'Anconna, 1954). Among many recent studies of the dynamics of food webs and food chains, largely through simulation by computers, a significant advance has been made by Smith (1969), who has incorporated the universally nonlinear response of individuals to their food supply. He finds that the fact that individuals can be satiated by food leads to an equilibrium distribution of matter or energy among the various species in a manner that depends largely on the input of matter or energy to the system and is little affected by the parameters of the components (evolved characteristics of the species). Through such approaches, it may become possible to predict the biomasses of plants, herbivores, and carnivores that a particular environment can support. The way in which these biomasses are divided up among species cannot, however, be fully understood from food-chain studies, and this problem has traditionally been approached as an extension of equations and models of competition. There is a large literature on competition and niche diversification, perhaps most thoroughly explored by MacArthur and his colleagues (e.g., MacArthur, 1968). Here too, the aim is prediction of the equilibrium abundances of various species in an environment.

In view of such great recent advances in understanding the control of whole systems and their component species, the chapters in this book, dealing largely with single-species populations, may seem to some readers to have a faintly old-fashioned flavor. Yet there is much to be learned from such studies, and certain problems can properly be pursued only with single-species populations. Certainly, the controversies involved are real ones.

Formally, as Lotka (1925) pointed out, it is simplest to begin by assuming an environment in which the population varies in size but everything else remains constant, so that the general equation 1 becomes

$$\frac{dN}{dt} = f(N) \tag{2}$$

Most controversies about the regulation of animal populations may be said to derive implicitly from this simplification and the forms of functions that are assumed. Such functions can be approximated to desired degree of accuracy by adding terms in an expanded series (Lotka, 1925),

$$\frac{dN}{dt} = a + bN + cN^2 + dN^3 \ldots, \tag{3}$$

where a must be zero, since we cannot have growth of N in the absence of N. This equation is to be viewed not only as an exercise in approximation since, from it, other more "theoretical" equations are derived. The very simplest case of this equation is the familiar one for exponential growth,

$$\frac{dN}{dt} = bN \tag{4}$$

Although simple, equation 4 rests on the enormous premises of an identical and unchanging character of all individuals in N, or at least on the existence of a "stable age distribution" in which the quality of individuals in each age group is unaffected by their abundance.

The term b is usually given as $r$ in the literature, and often called the "intrinsic rate of increase." There have been some terminological and conceptual confusions about this. Lotka (1925 and later works) conceived of $r$ as "true," "natural," or "intrinsic" partly in the sense of being given by the prevailing constant death and birth rates after the achievement of the appropriate stable age distribution. However, he also recognized that the maximum $r$ possible was intrinsic in the sense of being an inherent characteristic of a species. Andrewartha and Birch (1954) made this distinction clear by using $r_m$ (for $r_{max}$) and "innate capacity for increase," assumed to be unaffected by population size or shortage of resources, although influenced as a physiological variable by other aspects of the environment, and reserving $r$ for "actual rate of increase." It is perhaps unnecessary to dispense with Lotka's phrase, which has generally been used to connote maximum potential rate of increase. But regardless of the words used, failure to distinguish the concepts of $r$ and $r_{max}$ can lead to difficulties in interpreting the literature.

The problem in population regulation, then, is to understand why $r$ should over the long term generally average close to zero, even though $r_{max}$ is positive within the natural range of most species.

## Density-Dependence and Density-Independence

The first controversy in this book (as documented in the chapters by Solomon, Schwerdtfeger, Horn, and Tanner) is a classical one concerning fundamental assumptions about population growth. Probably in the eyes of many ecologists, this controversy has been resolved. Yet so much has been written on the subject by so many workers that the nature of the resolutions seems worth reviewing.

Much of the controversy derived from the belief that there is a deductive basis for the assumption that populations are "regulated" in numbers (by "mechanisms," in the words of Schwerdtfeger). Because many authors have denied this, it seems useful first to outline the essential arguments involved. The consequences of continually positive growth rates are, of course, unlimited population size; and of negative growth rates, ultimate extinction. But it is possible to consider the case where death and birth rates are nicely balanced over the long term except for purely random variations. The consequence of one simple form of random control, where there is equal probability of death or fission of each individual in each generation, is inexorable extinction (Skellam, 1955). Hutchinson (1961) has offered some perspective by calculating that the probability of such random extinction of a rare phytoplankter (1 cell per liter) in a lake of a million cubic meters is about 1 per cent after a million years. Other authors, like Schwerdtfeger (chapter 2 in this book), have examined the effects of "chance" variations in environmental influences on whole populations. Here it is necessary to define the variance of the random influences on population size, but again the probability of extinction is increasingly remote with larger populations. The time scales involved in random extinction would not seem to satisfy authors (e.g., Solomon, chapter 3 in this book) who would derive other kinds of control as an explanation of the long-term persistence of populations. Nor, however, does chance variation seem to be a sufficient explanation of the ultimate disappearance of all species (Andrewartha and Birch, 1954, pp. 663–65).

The proper deductive basis for assuming nonrandom control of populations (perhaps first noted by Haldane, 1953) seems to lie in natural selection, which will promote genotypes for relatively higher $r$ (not necessarily $r_{max}$) within the range of environmental variation to which the species is exposed. Random control itself cannot maintain $r = 0$ over the long term against the evolutionary adaptability of organisms, although it may do so in marginal or temporary environments supplied with genotypes evolved elsewhere. Therefore upper limits of population size must at least sometimes become imposed by something other than random events in the environment. Extinctions of once abundant species may be ascribed to "causes," such as nonrandom and perhaps rapid environmental trends, although chance extinction remains important in small subpopulations of species and may be fundamental in colonization.

Indeed, even those who stress the role of chance in population biology generally recognize, like Schwerdtfeger in his models of chance, that more regular restrictions on population growth may come into play at some times. Again, the essential arguments about the nature of these restrictions have been developed axiomatically and with simple equations by a number of authors. The "governor" of population growth, as pointed out by Cole (1957), cannot be a function of time alone (akin to ideas on "racial senescence" in older evolutionary literature) nor of population size alone, if we are to allow differences in the suitability of environments. Adding one more term from the expanded series of equation 3 to that for exponential increase in equation 4 gives a situation in which individuals may interact (as $N^2$) to affect population growth:

$$\frac{dN}{dt} = bN + cN^2 = r_{max}N + cN^2 \qquad (5a)$$

Here the value of c, which is negative for limited growth, may be determined by the environment, satisfying Cole's argument. This equation, which from its derivation is seen as the simplest possible imposition of a limit on exponential growth, is the familiar Pearl-Verhulst "logistic," perhaps more often seen as

$$\frac{dN}{dt} = r_{max}N \frac{(K-N)}{K} \qquad (5b)$$

where K is simply $N_{max}$, the equilibrium population size. Seen thus, the r of the population, or the specific growth rate, dN/Ndt, which it clearly becomes important to distinguish from $r_{max}$, is inversely proportional to population size.

The logistic may be understood in its most general sense as simply expressing the idea that conditions for population growth become increasingly inimical and restrict growth as the population increases, without specifying the mechanism except as "formally self-regulating" (Hutchinson, 1948). However, use of K rather than $N_{max}$ in the literature has perhaps served to emphasize the role of the environment. Thus K is often called "carrying capacity" of the environment, sometimes without explicit reference to the logistic, and (K − N)/K the "environmental resistance" to $r_{max}$.

The logistic has been a controversial equation, but there is probably little point in belaboring it as an admittedly unrealistic model of population growth. Attacks on its supposed mathematical shortcomings (Andrewartha and Birch, 1954, p. 411) seem to be based on misunderstandings (Philip, 1955). It is also possible to add terms to the logistic that are theoretically meaningful and improve its empirical fit; good examples are by Smith (1963), who used biomass rather than numbers and added a constant to allow for turnover time of the population at equilibrium, and by Williams (1967), who accounted for shortage of resources that result in rates smaller than $r_{max}$ at the outset of growth of populations of microorganisms. More serious shortcomings arise from the assumption of instantaneous response of $r$ to changes in N, and Wangersky and Cunningham (1957) and others since have shown how incorporation of time lags into the logistic and other equations greatly improves their accuracy.

The recurrent controversy has involved not the logistic but the broad concept it implies—not only that populations are ultimately limited in size by the inimical consequences of their own abundances, but also that regulation may operate on populations below this ultimate limit. It should be stressed that the controversies have not merely been restricted to the intraspecific competition subsumed by the logistic. Disease, parasitism, predation, and interspecific competition may all restrict population growth in ways that depend on the size of the population, but that require their own models and concepts, some of which are discussed here in chapter 1 by Solomon. Yet it is noteworthy that authors who have rejected the importance of intraspecific competition have often rejected all kinds of regulating influences related to population size.

The arguments have traditionally been couched in terms of density-dependence and density-independence. Some disputes have come from attempts to define these terms and to classify "factors" as density-dependent or density-independent. Much of the literature has been polemical, often confusing, and at times acrimonious, and it has therefore seemed best to introduce this subject with the short and excellent classification and resolution of terms and concepts by Solomon. There remains a core of respectable controversy about the relative importance of density-dependence and density-independence. It is totally impractical to include here any of the lengthy papers in which the more committed protagonists express their views. Solomon, rejecting weather as a sufficient control of populations, and Schwerdtfeger, stressing the role of chance, present partially opposed points of view fairly and reasonably. (It should be stressed in fairness to both authors, writing during the peak of this classical controversy, that they have since presented much more extensive and sophisticated accounts of population control—for example, Solomon in 1964, and Schwerdtfeger in 1968.) Differences among other authors in materials, concepts, and styles, as well as conclusions, can be much more extensively sampled in the pages of the 1957 *Cold Spring Harbor Symposia on Quantitative Biology* and in references listed in a number of the papers reprinted here.

Chapter 4 by Horn is included here as a succinct and elegant summing-up that also does much to resolve the controversy. Horn, using in effect a generalized form of the logistic, shows clearly how the two extreme views of population growth are not mutually exclusive and how both can be involved to varying degrees in determining the size of any population.

The controversy may have arisen partly from the kinds of organisms being studied. This probably should have been cleared up some years ago by Hutchinson's (1953) discussion of the relationships among life span, environmental fluctuations, and population control. His discussion is in the context of competition, but applies more generally to the tendency of populations to come to equilibrium through density-dependent growth. Organisms with very short life cycles (e.g., bacteria, protozoa) relative to the prevailing frequency of environmental fluctuations may experience favorable conditions for some generations before suffering from environmental change. Those with relatively long life cycles (e.g., vertebrates) must be resistant to the amplitude of changes that they encounter. Organisms like insects, with generations of the same order of frequency as environmental fluctuations, may be much influenced by changes from generation to generation and are likely to be reduced catastrophically before resources, competitors, or predators become restrictive. The controversy could not be entirely settled in this way, however, because many of the disputants have been entomologists.

A supposed conflict of viewpoint between "evolutionary" and "functional" ecologists as an explanation of controversies about population regulation (Orians, 1962) has been widely quoted in the literature. Careful reading of representative authors (see references to the evolutionary writings of the functional ecologists, Andrewartha, Birch, and Milne, in chapter 9 by Chitty in this book) suggests that the distinction may be invalid. The conflict of viewpoint may explain belief in competition as an explanation of differences between species (Birch and Erlich, 1967), but not apparently the importance that various authors attach to density-dependence. Horn uses evolutionary arguments in a quite different way to suggest resolution of the controversies. He suggests that species kept persistently below the level of equilibrium may be selected for high $r_{max}$, even at the expense of resistance to the environment, thus perpetuating their density-independent responses. The opposite may apply to organisms kept near equilibrium. Such divergent evolutionary tendencies may be partly responsible for the divergent points of view of students of the two sorts of populations.

The above discussion, and indeed much of the controversy, has been theoretical. Curiously, in all the years of dispute, it was only quite recently recognized that density-dependence is a statistically verifiable relationship. Tanner's recent and extensive survey has been chosen here to represent this empirical approach to the controversies (chapter 5). His chapter also confirms the general prevalence of density-dependence, supporting Horn's assertion that close association between numbers and environmental fluctuations does not necessarily preclude the effects of density. It should also be noted that there are certain statistical and analytical controversies in this approach to demonstrating density-dependence, which can be traced through the recent note by Morris and Royama (1969).

Possibly some of the most promising approaches to finding causes of population control are also statistical, through demonstrations of key factors. A good review is by Solomon (1964). Yet generalizations and models will continue to be sought, and controversies about causality can be extended to every individual case. The later chapters in this book are concerned with multifarious causes of population control.

## SOCIAL REGULATION OF POPULATIONS

The above discussion emphasized extrinsic sources of regulation of populations — weather, food shortages, and the like. There is increasing evidence that density-dependent control can result from behavioral interaction of individuals or groups, rather than from negative responses of the environment. The reality of such regulation* by populations of some species is indisputable, but it has given rise to a number of controversies about mechanisms, the general prevalence of such regulation in nature, and its function and evolutionary origin. Most forms of social regulations are discussed or at least mentioned here in the chapters by Christian and Davis, Chitty, Wynne-Edwards, and Wiens. Because of the varied contexts in which these authors discuss their ideas, sometimes to support conflicting points of view, it seems useful to attempt a brief classification of the whole field of social regulation as a framework in which the controversies can be discussed.

Although most study has concerned the negative effects of crowding and behavioral interactions, cases where $r$ is a positive function of N (confusingly referred to as inverse or positive density-dependence by some authors — see Solomon's discussion in chapter 1) are not explored in this book, but deserve brief consideration. Of course the relationship between $r$ and N may

* The term "self-regulation," as used by authors in this book, has appeared in the literature to mean any sort of density-dependent growth, and its application in the narrower sense of excluding extrinsic restrictions has been disputed (several discussants, in LeCren and Holdgate, 1962, pp. 378–80). The term "social regulation" used here might, on the other hand, require rather broad meaning of the word "social" in such cases as autoxicity by metabolic products and some other cases of greater natural significance as discussed in this section.

be positive in small populations (and hence may threaten extinctions) for statistical rather than behavioral reasons, where the chances that mates will meet are reduced (Lotka, 1925; Haldane, 1953; and others). In fact, Philip (1957) defined the most elementary expression of social behavior as nonrandom movement of individuals, resulting in enhanced $r$ in rarified populations. With the possible exception of cases of conditioning of the environment, other positive effects of N on $r$ appear to be based on groupings of individuals. These may be only vaguely social, like the possible effects of clumping in space in restricting the chances of discovery by predators. A similar strategy in time may involve little social interaction (as in the periodic saturation of the environment by cicadas — see Lloyd, 1966) or may involve social facilitation of synchronous breeding by birds to satiate nest predators (Tinbergen, 1965). However, schooling and social breeding may have much more obvious benefits through enhanced feeding or defense by individual participants (Tinbergen, 1965). More conspicuous benefits of truly social groupings may occur where genetic kinship is involved, as in families or insect societies. In spite of many striking examples of positive relationships between $r$ and N, ideas on "proto-cooperation" and the universality of positive effects of social interactions (Allee, 1931) seem to reduce to special cases. It should be borne in mind that, in spite of the advantages of such groupings, behavior may also impose equilibrium within the group when the group becomes large; it is this negative effect of N on $r$ which is the "advantage" postulated by Wynne-Edwards (see discussion below).

To introduce such negative effects of behavior on population growth, we next turn to one kind of behavior that may impose restrictions through intensified use of resources without, *in theory*, involving social interactions between individuals. Although homing and tradition have been studied extensively, their effect in limiting population growth has received little attention. The only general discussion of this effect known to me is by Snyder (1948), although it should be noted that Wynne-Edwards sees such behavior as evolved for the purpose of limiting population growth socially (see below). An orthodox general explanation for homing and traditional use of sites would be that places suitable for survival or reproduction in the past — as measured by the prevalence of descendents of users of these places — will more likely be suitable in the future. Exceptions among colonizing or "fugitive" species (Hutchinson, 1951) prove the rule, since past habitats of these are predictably unsuitable. Even under conditions of crowding, where resources may become limiting, the population may continue to be largely recruited from those individuals that are *relatively* successful in traditional sites, thus perpetuating selection for site faithfulness or homing, restricting population growth locally, and inhibiting geographical spread.

Other forms of social regulation of populations may be effective, *in theory*, even where food is in excess of that required for higher *r*. The simplest form is perhaps defense of sites or home ranges by territorial species. Where suitable sites or potential home ranges are limited in number, they might be thought of as "resources" that are increasingly difficult to discover or defend as populations increase. Territoriality certainly occurs in mammals, but it may be that social stress induced by competition for sites and territories functions more directly in imposing equilibrium on their populations (see below). On the other hand, the proportionate relationship between metabolic demand (assumed to be proportional to weight to about the $\frac{2}{3}$ power) and area of home ranges of various species of wild mammals (McNab, 1963) suggests that if mammals in general defend their home ranges, this is for the good reason that food resources might otherwise restrict the defender's reproductive output or survival. In this case, it might be more proper to think of food rather than behavior as the controlling factor. A number of adaptive functions of defense of home ranges by birds have been suggested. Armstrong (1965) concluded from extensive examples that birds in general have larger territories (exclusive home ranges) than dictated by their metabolic demand. Schoener (1968) recently analyzed more material and came to opposite conclusions by selecting, grouping, and correcting data in ways that seem logically defensible, but that may nevertheless involve an element of "significance seeking." Although the regulating effect of territory on population size is widely assumed (see chapters 7, 8, and 9 by Wynne-Edwards, Wiens, and Chitty), it appears that more empirical study is required.

Most remaining forms of social regulation of animal populations might be thought of as resulting from numbers of "harmful contacts" between individuals, although these may be a consequence of defense of sites or territories. This behavior can have the effect of keeping a population below the level permitted by available resources or by possible requirements for territories or sites. A simple model of such regulation is the logistic in the form of equation 5a, where c is taken as a coefficient of behavioral interaction rather than as a property of the environment. Other more complex modifications of the logistic as a model of social regulation by groups of animals are outlined by Slobodkin (1953), and certain time lags in population response (Wangersky and Cunningham, 1957) may have particular relevance in this form of regulation (see chapter 9).

Cannibalism may perhaps be viewed as a form of harmful contact that can lead to population regulation. Cannibalism has been much studied in the laboratory (see examples in chapter 7), where its occurrence is by definition unnatural. In nature, the regulating role of cannibalism among fish may be important, but, interestingly, cannibalism may also be necessary for maintenance of single-species populations where young eat food inaccessible to or inadequate for larger fish, which in turn depend on cannibalism for growth and even maturation (e.g., Patriquin, 1968).

Remaining forms of social regulation are less stark and much more widespread. In such populations, crowding is as likely to result in impaired or thwarted reproduction as in increased mortality. As might be expected, such population restrictions have been found most widely among higher animals with well-developed social behavior. Indeed, mammals are the sole concern of chapter 6 by Christian and Davis, chosen to represent the most extensive development of ideas on social regulation through "harmful contacts." Such work presents an array of problems for physiologists and behaviorists. To Christian and Davis, the mechanisms are centered in adrenocortical physiology. The evidence for this is controversial, especially in nature, and Christian and Davis deal at length with questions of seasonal cycles and sexual cycles, which may seem to give the theory a somewhat Ptolemaic character, but which are fundamental. Chitty, in chapter 9, does not reject the importance of stress, but considers it inadequate to account for numerical changes in populations of small mammals.

No doubt there will continue to be controversies about mechanisms and the general prevalence of the many forms of social regulation outlined above. But far more vigorous disputes have arisen about the "purpose" of such regulation. An orthodox statement about the evolution of social regulation can be found in the closing words of an earlier essay by Christian (1961): "Since the survival and evolution of a species must depend on reproduction by dominant individuals, the selective process would seem to be operating in a direction to increase the importance of behavioral adaptation." Yet it is easy to ascribe a function to such regulation. A circumspect statement at the end of the chapter by Christian and Davis suggests that their behavioral-endocrine "feedback mechanism acts as a safety device, preventing utter destruction of the environment and consequent extinction." Teleological statements in biology have a certain value, provided they are understood as succinct ways of expressing the necessity of a particular structure or function for the survival of an individual, or of describing the adaptive consequences of evolution. Therefore there is a semblance of respectability in the suggestions of a number of authors that social regulation is for the "purpose" of preventing overutilization of resources and consequent dangers of reduced production or even extinction of the population.

Before going on to consider such suggestions, it should be pointed out that the premise of extreme overutilization of resources is not generally derivable from the writings of ecologists who have been concerned with the extrinsic regulation of populations, as discussed earlier. Some dangers might seem implicit in prey-predator interactions, but these in simplest interpretations lead to cycles rather than extinctions. Extinctions may occur in laboratory populations, but the existence of "refugia" and dispersal lags in complex laboratory systems (Huffaker, 1958), and perforce in nature, can maintain intrinsically unstable prey-predator interactions. Furthermore, interactions in complex food webs in nature (mentioned earlier, but outside the scope of this book) are widely believed to impose stability on component species. The idea of population homeostasis hardly depends upon the assumption of social regulation *

The most noteworthy proponent of the idea that social regulation is for the "purpose" of preventing overutilization of resources is Wynne-Edwards, who in chapter 7 gives the essence of his beliefs. Although the reality of social regulation had been well established in many earlier studies, Wynne-Edwards has crystallized the controversies. These have arisen, first of all, because he discusses many forms of behavior which had been given other explanations, but which he believes are involved in population control, for which he offers little or no direct evidence. Even more disputed have been his views on the evolution of such behavior. He attaches little importance to the adaptiveness to individuals of the several forms of behavior that may lead to social regulation in ways discussed above, and invokes group selection to explain them as benefits to the population or species. The concept of group selection is not unique to Wynne-Edwards' writings. This "heretical" idea and its expression in the writings of many biologists is discussed and documented further by Williams (1970) in a companion volume to the present one. The short chapter by Wiens deals excellently with the specific controversies raised by Wynne-Edwards' views. Wiens, in rejecting Wynne-Edwards' arguments about group selection, believes in the importance of

---

* It is worth noting that one frequently cited textbook example of overutilization of resources—that by deer of the Kaibab Plateau after removal of predators—has probably been misinterpreted and overstated. Howard (comment on p. 483 in Baker and Stebbins, 1965) indicates that fire and overgrazing by sheep were also involved.

tenuous effects of individual selection, arguing that there may be certain values in hierarchical and dominance behavior even for subordinate individuals, which by submitting, may gain later opportunities. Chitty, in chapter 9, without considering Wynne-Edwards' views at length, rejects them by stressing the relative nature of genetic fitness. There is no need, Chitty argues, to assume that such behavior is adaptive even for successful individuals; the dominant individual may prevail by suppressing others even if his own reproductive output is somewhat (although less) impaired. Wiens charitably concludes that Wynne-Edwards' ideas on social regulation should not be discarded because of the distractions posed by arguments about evolution. Certainly Wynne-Edwards' work has stimulated wide discussion and re-assessment of established viewpoints.

## Genetic Aspects of Population Regulation

It is appropriate to turn from a controversy about the evolution of behavior to the larger questions of the role of genetic change and natural selection in the regulation of populations in general.

  We consider first those aspects of natural regulation that Lack (1965) calls "evolutionary ecology." As in other fields of biology, there has been increasing recognition that the mechanistic or functional interpretations which have been paramount since the end of the nineteenth century are incomplete without the evolutionary viewpoint. It can be asserted that vital parameters — death rates, birth rates, maturation rates — must be partly inherent, like any other features of organic existence. The growth and regulation of animal populations, even when thought to be "imposed" by the environment, must therefore contain an intrinsic, genetically controlled component representing the adaptive outcome of natural selection. The influence of such intrinsic components is evident in a naive way in deriving the logistic (see equations 5a and 5b) where the reader may satisfy himself that K is a substitute for $r_{max}/c$; equilibrium population size (K) is not only determined by the environment through c, but by the (inherent) intrinsic rate of increase, $r_{max}$. The seminal paper on such subjects is clearly that by Cole (1954), who compared effects on intrinsic rate of increase of differences in age of maturity, successive reproductive schedules, fecundity, and longevity, and thus set the stage for studies of the actual demographic strategies practiced by various species. Although many organisms seem to depend on high intrinsic rate of increase, there are constraints leading to a whole array of strategies. First, especially among poikilotherms, there may be strong physiological links among vital parameters, so that, for example, accelerating maturity reduces fecundity (McLaren, 1963), or reducing size enhances $r_{max}$ but increases mortality hazards (Smith, 1954). Second, the magnitude of vital parameters must be understood as resulting from opposing selective forces. The best-known example is clutch size of some birds, which is thought to be determined by the number of young that can be raised successfully (Lack, 1965). A more complex example (Brooks, 1965) are the cyclomorphic *Daphnia* in lakes, which have forms in summer that appear to "waste" energy (perhaps equal to one half egg per instar) by producing elongate heads (helmets). Helmets, however, apparently confer advantage against increased predation by fish in summer. Third, most organisms have to contend with temporally variable environments; and the strategies of repeated reproduction, which may not increase $r_{max}$, must be understood in relation to this (e.g., Murphay, 1968). Also, the strategies of potential rate of increase may vary with amount of density-independent mortality or degree of resource limitation (Horn, chapter 4). All such problems can generally be understood if the differences between $r$ and $r_{max}$ are kept in mind; the latter may be reduced to maximize the former. Some very sophisticated analyses of demographic strategies are being developed by a number of workers, and the whole field of inquiry will doubtless become more active. One promising aspect is the possibility of translation to measures of fitness in population genetics.

If we accept the above arguments that vital parameters are the result of natural selection, then we must accept the possibility of evolutionary change. More controversies in population ecology have arisen from this dynamic aspect of adaptation. Although population geneticists have accepted for some time the possibility of rapid evolutionary change, only recently has this been entertained by population ecologists. The controversies arise not from theoretical hesitancy about the relevance of change but from our present ignorance about the magnitude and general prevalence of such change in short time periods.

Two authors have been prominent in developing ideas about genetic selection and population regulation. Pimentel, in chapter 10, sees short-term genetic changes in predator efficiency, prey vulnerability, virulence of parasites and pathogens, host resistance, and competitive abilities, all in response to any imbalance or eruption that might occur. The effect is to maintain homeostasis in spite of conflicting strategies. Chitty, on the other hand, derives his views as expressed in this book by assuming widespread operation of social regulation. This being so, important variations in selective forces may derive from the size of the population itself and the associated changes in behavior of individuals. As Chitty argues, this may lead to oscillations, through mechanisms summarized on his Figure 3, rather than to steady population equilibrium.

In assessing these two points of view, one is struck by certain difficulties in testing Pimentel's hypothesis. In stable associations of species, evolution is presumably reflected only in the established adaptations, and need not be part of the mechanistic interpretation of population regulation. His experimental studies, on the other hand, may involve selection pressures larger than those occurring in nature. He cites a number of examples of genetic change following introductions and outbreaks, yet the history of extinctions and permanent restrictions of numbers or range following such events (Elton, 1958) does not seem to manifest widespread potential for rapid evolutionary adjustment. It might be argued that in these cases the advent of man has introduced selective forces to which organisms have not been "preadapted," but this leaves us with the difficulty of assessing the significance of genetic change in near-stable systems.

Chitty's views, on the other hand, seem much more testable

for populations that show natural fluctuations in ways which he outlines. Although he does not say so explicitly, much of the genetic feedback involved in his hypothesis would be mediated by sexual rather than natural selection. Sexual selection, acting on individuals of maximum reproductive value (i.e., newly matured males) and involving very large differences in fitness (between breeding, perhaps polygynously, and not reproducing at all) might lead to much more rapid evolutionary changes than the natural selection considered by Pimentel. Even so, some workers have hesitated to accept the rates implied by Chitty's hypothesis (see, for example, the comments by Christian and Davis). Theoretical enquiries into the feasibility of rapid evolutionary changes in polygenic traits like body size, aggressiveness, and fertility cannot be made using the approaches of classical population genetics. However, heritability coefficients of such traits and the associated density-dependent fitnesses could give some measure of evolutionary rates, at least over periods of a few generations (Lee and Parsons, 1969).

If there are further controversies in Chitty's ideas, these may derive from an underestimation of the nongenetic component in inherent changes of quality of individuals. This point is made by Christian and Davis, although they offer no direct evidence on the heritability of the stressed state beyond a generation. Prenatal influences, however, do occur widely. A pertinent recent study by Bryden (1968) showed that changes in body size, thought by Chitty to be important among small mammals, may be transmitted phenotypically through successive generations of seals, and are thought to be involved in social regulation of their populations.

We may certainly look forward to increased attention to the ideas of both Pimentel and Chitty, but I believe that genetic feedback will have more relevance in populations that show social regulation.

## REFERENCES

Allee, W. C. 1931. *Animal aggregations. A study in general sociology.* University of Chicago Press, Chicago. 431 pp.

Andrewartha, H. G., and L. C. Birch. 1954. *The distribution and abundance of animals.* University of Chicago Press, Chicago. 782 pp.

Armstrong, J. T. 1965. Breeding home range in the nighthawk and other birds; its evolutionary and ecological significance. *Ecology*, 46:619–629.

Baker, H. G., and G. L. Stebbins. 1965. The genetics of colonizing species. *Proc. First Int. Union Biol. Sci. Symp. Gen. Biol.* Academic Press, New York. 588 pp.

Birch, L. C., and P. R. Ehrlich. 1967. Evolutionary history and population biology. *Nature*, 214:349–352.

Brooks, J. L. 1965. Predation and helmet size in cyclomorphic. *Daphnia. Proc. Nat. Acad. Sci. Wash.*, 53:119–126.

Bryden, M. M. 1968. Control of growth in two populations of elephant seals. *Nature*, 217:1106–1108.

Christian, J. J. 1961. Phenomena associated with population density. *Proc. Natl. Acad. Sci. Wash.*, 428–449.

Cole, L. C. 1954. Population consequences of life history phenomena. *Quart. Rev. Biol.*, 29:103–137.

———. 1957. Sketches of general and comparative demography. *Cold Spring Harbor Symp. Quant. Biol.*, 22:1–15.

D'Anconna, U. 1954. *The struggle for existence.* E. J. Brill, Leiden. 274 pp.

Elton, C. S. 1958. *The ecology of invasions by animals and plants.* Methuen, London. 181 pp.

Haldane, J. B. S. 1953. Animal populations and their regulation. *New Biol.*, 15:9–24.

Huffaker, C. B. 1958. Experimental studies on predation: dispersion factors and prey-predator oscillations. *Hildigardia*, 27:343–383.

Hutchinson, G. E. 1948. Circular causal systems in ecology. *Ann. N.Y. Acad. Sci.*, 50:221–246.

———. 1951. Copepodology for the ornithologist. *Ecology*, 32:571–577.

———. 1953. The concept of pattern in ecology. *Proc. Acad. Nat. Sci. Philadelphia*, 105:1–12.

———. 1961. The paradox of the plankton. *Amer. Nat.*, 95:137–145.

Lack, D. 1965. Evolutionary ecology. *J. Anim. Ecol.*, 34:223–231.

LeCren, E. D., and M. W. Holdgate, eds. 1962. *The exploitation of natural animal populations.* Blackwell, Oxford. 399 pp.

Lee, B. T. O., and P. A. Parsons. 1969. Selection, prediction and response *Biol. Rev.*, 43:139–174.

Lloyd, M. 1966. The periodical cicada problem. I. Population ecology. *Evolution*, 20:133–149.

Lotka, A. J. 1925. *Elements of physical biology.* Williams & Wilkins, Baltimore. 460 pp.

MacArthur, R. H. 1968. The theory of the niche. In R. C. Lewontin, ed. *Population Biology and Evolution.* Syracuse University Press, Syracuse, N.Y. Pp. 159–176.

McLaren, I. A. 1963. Effects of temperature on growth of zooplankton and the adaptive value of vertical migration. *J. Fish. Res. Bd. Canada*, 20:685–727.

McNab, B. K. 1963. Bioenergetics and the determination of home range size. *Am. Nat.*, 97:133–140.

Morris, R. F., and T. Royama. 1969. Logarithmic regression as an index of responses to population density. Comment on a paper by M. P. Hassell and C. B. Huffaker. *Can. Entom.*, 101:361–364.

Murphay, G. I. 1968. Pattern in life history and the environment. *Am. Nat.*, 102:391–403.

Orians, G. H. 1962. Natural selection and ecological theory. *Am. Nat.*, 96:257–263.

Patriquin, D. G. 1968. Biology of *Gadus morhua* in Ogac Lake, a landlocked fiord on Baffin Island. *J. Fish. Res. Bd. Canada*, 24:2573–2594.

Philip, J. R. 1955. Note on the mathematical theory of populations dynamics and a recent fallacy. *Austr. J. Zool.*, 3:287–294.

———. 1957. Sociality and sparse populations. *Ecology*, 38:107–111.

Schoener, T. W. 1968. Sizes of feeding territories among birds. *Ecology*, 49:123–141.

Schwerdtfeger, F. 1968. Eine intergreite Theorie zur Abundanz-dynamik tierischer Populationen. *Oecologia* (Berlin), 1:265–295.

Skellam, J. G. 1955. The mathematical approach to population dynamics. In J. G. Cragg and N. W. Pirie, eds., *The numbers of men and animals*. Oliver and Boyd, Edinburgh. Pp. 31–45.

Slobodkin, L. B. 1953. On social single species populations. *Ecology*, 34:430–434.

Smith, F. E. 1954. Quantitative aspects of population growth. In E. Boell, ed., *Dynamics of growth processes*. Princeton University Press, Princeton, N.J. Pp. 277–294.

———. 1963. Population dynamics in *Daphnia magna* and a new model for population growth. *Ecology*, 44:651–663.

———. 1969. Effects of enrichment in mathematical models. In *Euthrophication: Causes, consequences, corrections*. National Academy of Sciences, Washington, D.C., pp. 631–645.

Snyder, L. L. 1948. Tradition in bird life. *Canadian Field-Nat.*, 62:75–77.

Solomon, M. E. 1964. Analysis of processes involved in the natural control of insects. *Adv. Ecol. Res.*, 2:1–58.

Tinbergen, N. 1965. Behavior and natural selection. In J. A. Moore, ed., *Ideas in modern biology. Proc. V. 6, XVI. Int. Congr. Zool.* Pp. 521–542.

Wangersky, P. J., and W. J. Cunningham. 1957. Time lag in population models. *Cold Spring Harbor Symp. Quant. Biol.*, 22:329–338.

Williams, F. M. 1967. A model of cell growth dynamics. *J. Theoret. Biol.*, 15:190–207.

Williams, G. C., ed. 1971. *Group selection*. Atherton, New York. In press.

# 1. Meaning of Density-Dependence and Related Terms in Population Dynamics

## M. E. SOLOMON

One of the most familiar concepts of population dynamics is the idea of the regulation of abundance by density-dependent factors, which operate more severely at high density than at low density, per individual of a population. But there has been some variation in the meaning and scope of "density-dependent" and related terms as used by different authors, and in recent years this tendency has increased. This has led to some confusion, and poses the question how the application of these terms can be stabilized. Varley[1-3] has argued that the meaning and scope of such a term should be determined strictly according to its author's definition, and that on this basis insect parasites, predators, and disease organisms should be excluded from the density-dependent category. Discussion of this matter with Prof. Varley has not altered my opinion that it is too late to attempt to make a change so contrary to the intentions of the

From *Nature*, 181 (June 28, 1958). 1778–1780.

originator of this term, and to subsequent common usage, but has led to various improvements in the present article.

In order to deal with a number of points which are closely interrelated, it is proposed: (1) to look back to the origins of the terms concerned, then (2) to give a brief account, similar in some respects to that given by Varley,[3] of the variations in the use of these terms by various authors, and of a number of synonyms, and finally (3) to discuss various proposals for clarifying the position.

(1) Howard and Fiske,[4] discussing the natural control of insects with special reference to defoliating caterpillars, wrote that for natural control, or balance, to exist "it is necessary that among the factors which work together in restricting the multiplication of the species there shall be at least one, if not more, which is what is here termed facultative . . . , and which, by exerting a restraining influence which is relatively more effective when other conditions favor undue increase, serves to prevent it." The chief of such facultative factors was stated to be parasitism, "which in the majority of instances, though not in all, is truly 'facultative.'"

They considered disease as a type of facultative factor which, in their experience, became effective chiefly at a very high level of abundance, and starvation as operating only at the extreme limit. Generally, insects would be controlled at a low level of density "only through parasites and predators, the numerical increase of which is directly affected by the numerical increase of the insect upon which they prey."

Secondly: "A very large proportion of the controlling agencies, such as the destruction wrought by storm, low or high temperature, or other climatic conditions, is to be classed as catastrophic, since they are wholly independent in their activities upon whether the insect which incidentally suffers is rare or abundant. . . . The average percentage of destruction remains the same. . . ."

Thirdly: "Destruction through certain other agencies, notably by birds and other predators, works in a radically different manner . . . it may be considered that they average to destroy a certain gross number of individuals each year, and . . . this destruction . . . would most probably represent a higher percentage when that insect was scarce than when it was common. In other words, they work in a manner which is the opposite of 'facultative' as it is understood."

In short, Howard and Fiske recognized three types of factors: (*a*) "catastrophic"; (*b*) "facultative" and a third type which was later called (*c*) "inverse."[5] Later, Thompson[6] divided controlling factors into two classes. One, which he named "general," was the same as Howard and Fiske's catastrophic type. The other comprised "those *individualized* factors whose destructive capacity depends in some way on the numerical value of the host population;" this definition takes in two of Howard and Fiske's categories, the facultative and the inverse factors; it is clear from this paper and from Thompson's later writings[7] that he means chiefly parasites, but he considers that, because of their limited reproductive capacities, they generally act as inverse factors.

It was H. S. Smith[8] who introduced the new terms "density-independent" and "density-dependent" for Howard and Fiske's catastrophic and facultative categories. He made the mistake of identifying the facultative or density-dependent class with Thompson's individualized category; but this designation, "destroying a percentage which increased when the numbers of the host increased," shows that he did not mean to include inverse factors; indeed, he followed Howard and Fiske in recognizing them as a distinct third category, again citing "many insectivorous birds and mammals" as examples. His examples of density-independent agents included climatic factors, "intrinsic" mortality, and malnutrition due to unsuitable food. Examples of density-dependent factors were, primarily, entomophagous insects, then certain infectious and contagious diseases, quantity of food and competition for nesting sites or protective niches; the last-mentioned process, it was suggested, was a medium through which climatic factors could have a density-dependent action.

Thus Smith and Howard and Fiske agreed in designating facultative or density-dependent factors as those having a greater proportionate influence at high than at low population density, and also in regarding insect parasites, or entomophagous insects

(Smith), certain pathogens, and food shortage as examples; Smith added intraspecific competition, and hence (in his view) certain climatic factors. Varley[1-3] has insisted that the definition and the examples are at variance, inasmuch as theoretically constructed parasite-host oscillations show a relationship which, owing to the delayed response of parasite numbers to changes in host numbers, is alternately density-dependent and inverse. It may be added that interactions broadly similar to the theoretical ones have been demonstrated in laboratory experiments and in a fragmentary form in certain field populations, though not in others. Nevertheless, the great majority of writers on the subject have accepted Smith's ideas and terms, presented as they were without formal definition, and have been content to use them in a not too precise way; in particular, most writers have accepted all those insect parasites and predators exerting a regulating influence upon a population as density-dependent factors; whether or not their response to host density might be so delayed as to prevent any consistent and straightforward density-dependence has been little considered.

(2) There have, however, been a number of variations in the application of Smith's terms, and various new terms have been used by authors wishing to avoid confusion. The term "density-independent" has suffered relatively little misuse. But in his textbook, Odum[9] refers to a type of inverse factor as density-independent, and uses a new term, "density-proportional," for one type of density-independent factor.

With respect to the term "density-dependent," a major source of confusion is that some authors have taken it to include inverse factors, which Smith specifically excluded. Allee[10] and Allee et al.[11] did this, and then separated "direct density-dependent" from "inverse density-dependent" factors. Haldane[12] did likewise, but called these two subdivisions "negative" and "positive density-dependent." Thompson[7] writes that Smith proposed the new term "density-dependent" for Thompson's individualized category. This, as already noted, was an error on the part of Smith;[8] for, like Howard and Fiske,[4] he explicitly separated the inverse type of factor into a separate category.

Milne[13] makes a distinction between "perfectly" and "imperfectly" density-dependent factors: "Perfection here means an *exact* linear (or curvilinear) relationship between increasing action of the factor and increasing density of the species." The only perfectly density-dependent factor is said to be competition between members of the population. I do not believe that factors perfectly dependent on density exist except at a high level of abstraction. In nature, the actions of all factors are presumably variable and inexact.

Varley,[1, 2] in separating parasites, predators and pathogens from the density-dependent category, called them "delayed density-dependent" factors. Lack[14] has treated this new category as a subdivision of the density-dependent class.

Some synonyms of density-dependent are "facultative,"[4] "controlling,"[15] "concurrent,"[5] and "density-governing."[16]

"Inverse" factors of various types have been called "density-disturbing,"[16] "inverse density-dependent,"[10, 11] and "positive density-dependent."[12] Odum[9] refers to one sort of inverse action as an inverse type of density-dependence, and to another sort as density-independent. The terms positive or inverse density-dependent, as we have seen, extend the density-dependent category beyond the limits laid down by its originator.

There is, of course, an important characteristic held in common by density-dependent and inverse factors: the action of both is related to population density, though in opposite senses. Terms (other than density-dependent) which have been used for including both these types of factors together are "individualized,"[6] "density-related,"[5] and "reactive."[16]

(3) Turning now to the question how to standardize the use of the term density-dependent and reduce the confusion which has beset it in the past, there are three main possibilities:

(*a*) The first is to follow the course advocated by Varley,[2, 3] and place a strict interpretation on the explanations (scarcely formal definitions) given by Howard and Fiske[4] and by Smith.[8] What the result of this would be depends upon whether or not many insect parasites and predators do in fact maintain an oscillatory relationship, of the type seen in mathematical theories, with the populations of their hosts or prey. If Varley's[2] expectations are correct, parasites and predators would generally be excluded from the density-dependent category of which Smith took them to be the prime exemplars. Besides running counter to the intentions of the originator of the term, this would give density-dependence a far narrower meaning than it has been given throughout the extensive literature in which it has been used over the past two decades. (The only exceptions to this, so far as I am aware, are papers by Varley.[1-3]) I believe such a restriction of the scope of this familiar term would be unlikely to find general acceptance, and that after such a long period it would cause a great deal of further confusion if it were adopted.

(b) The second possibility is to follow the example of Allee,[10] Allee et al.,[11] and Thompson,[7] and take the density-dependent category to include, not only all the types of factors attributed to it by Smith, but also the inverse factors, which he excluded. In support of this course, it may be said that it has been fairly widely favored, and that it interprets density-dependent in the literal sense of "acting in a way which is dependent on density." But I doubt whether it accords with a majority viewpoint, and am opposed to it because it gives the term density-dependent a quite different scope from that laid down by its originator.

(c) The third possibility is to follow the intentions of Smith as closely as possible, by including, along with intraspecific competition and other straightforward density-dependent factors, those parasites, predators, and pathogens which respond to changes in host population density in such a way as tends to limit these changes (even if the response is delayed as in the theoretical parasite/host oscillations), and by excluding only those factors the action of which wholly or predominantly maintains an inverse relationship with density, and those which are independent of density. Many writers have explicitly or implicitly adopted a similar interpretation. I suggest that, insofar as the term "density-dependent" continues in use, this is the way in which it should properly be applied.

Whichever course is followed, one major difficulty is inescapable for the present: there are very few ecological situations which are well enough known for particular biological factors to be placed with confidence in any one of Smith's three categories. Still less can we say in general what proportion of insect parasites and predators act as density-dependent, or inverse, or density-independent factors.

The writer on this subject may often prefer to avoid using the term "density-dependent." One alternative is to use Nicholson's term "density-governing." Nicholson[16] writes that "its meaning is precisely that of 'density-dependent factor' according to definition (Smith, 1935)," and, like Smith, he puts inverse factors in a separate category (density-disturbing). A possible objection to the term "density-governing" is the implication that all density-dependent factors have a governing influence upon density, which seems unlikely.

As a means of avoiding the various difficulties associated with the use of "density-dependent," and at the same time recognizing the special features of the factors Varley[1] called "delayed density dependent," the following classification may be useful. It sets out some types of relation which the unfavorable action of factors upon a population may bear to the density of that population, the action being measured as an effect upon mortality, the reproductive rate, or the net rate of increase or decrease, and expressed in proportionate terms such as percentage.

(1) "Directly density-related:" showing a positive correlation between adverse action and density.

(2) "Inversely density-related" (or "inverse"): showing a negative correlation between adverse action and density.

(3) "Alternately density-related" (or "alternating"): where a lagging response to changes in density leads to a regular succession of directly and inversely related phases.

This last category is distinguished from the density-independent (or nonreactive[16]) relationship, where the correlation between action and density would generally vary (in successive short runs of observations) in an irregular manner between positive, zero, and negative, with a long-run correlation not differing significantly from zero.

A single factor might fall in different categories in different circumstances, or at different levels of abundance or (as pointed out to me by Dr. R. F. Morris, of Canada) within a single generation as compared with a series of generations.

Upon such questions as what sort of ecological causes may lie behind the correlations observed, which factors may be found to fall in which category under various conditions, and what may be their function in natural control, this classification is deliberately left noncommittal.

## R E F E R E N C E S

1. Varley, G. C., *J. Anim. Ecol.*, 16, 139 (1947).
2. ———, Trans. Ninth Int. Congr. Entom., 2, 210 (1953).
3. ———, *J. Ecol.*, 45, 639 (1957); *J. Anim. Ecol.*, 26, 251 (1957).
4. Howard, L. O., and Fiske, W. M., *Bull. U.S. Bur. Ent.*, No. 91 (1911).
5. Solomon, M. E., *J. Anim. Ecol.*, 18, 1 (1949).
6. Thompson, W. R., *Parasitol.*, 20, 90 (1928).
7. ———, *Canad. Entom.*, 87, 264 (1955).
8. Smith, H. S., *J. Econ. Entom.*, 28, 873 (1935).
9. Odum, E. P., "Fundamentals of Ecology" (W. B. Saunders Co., Philadelphia and London, 1953).
10. Allee, W. C., *Amer. Nat.*, 75, 473 (1941).
11. ———, Emerson, A. E., Park, O., Park, T., and Schmidt, K. P., "Principles of Animal Ecology" (W. B. Saunders Co., Philadelphia and London, 1949).
12. Haldane, J. B. S., "New Biology," No. 15, 9 (Penguin Books, Ltd., London, 1953).
13. Milne, A., *Canad. Entom.*, 89, 193 (1957).
14. Lack, D., "The Natural Regulation of Animal Numbers" (Clarendon Press, Oxford, 1954).
15. Nicholson, A. J., *J. Anim. Ecol.*, 2, 132 (1933).
16. ———, *Austr. J. Zool.*, 2, 9 (1954).

# 2: Is the Density of Animal Populations Regulated by Mechanisms or by Chance?

## F. SCHWERDTFEGER

*Regulating* means the maintaining of a more or less varying, but on the average equal population density; in other words, the regulating influences cause the maintenance of a balance, which may be called a population balance if the equalizing of fertility and mortality within a population is meant, or a biocenosis balance in case one means the relation of numbers or of the mutual influences between the population and the other components of the biocenosis. Regulating factors become effective by depressing the density when it increases and by raising it when the abundance decreases.

Based on theoretical considerations which are supported by observations in the field and by experiments, many authors generally accept the possibility that such a regulation is bound to the presence of a mechanism, in such a manner that checking influences automatically become more effective when the

From *Proceedings of the 10th International Congress of Entomology.* 4 (1958). 115–122.

abundance increases and vice versa. As an example, Nicholson (1933, p. 135) writes: "For the production of balance, it is essential that a controlling factor should act more severely against an average individual when the density of animals is high, and less severely when the density is low. In other words, the action of the controlling factor must be governed by the density of the population controlled." Only density-dependent factors — in the broadest sense — are able to cause a balance. Influences which are independent of abundance, e.g., components of the weather, can considerably increase or decrease the density, but not in a determined manner; occasionally, they may become effective as regulators and thus assist or even replace the density-dependent or density-governed factors, but in principle the maintenance of balance will be caused by the latter.

It is proved in many cases that mechanisms regulate abundance and restore disturbed balances. The gradations of the pine noctuid, *Panolis flammea* (Schiff.), are regularly followed by an increase of the tachinid, *Ernestia rudis* (Fall.), which is promoted by the enlarged host supply. The parasite reduces the host again to a normal density; that is to say, to an average abundance that is inconspicuous for many years.

Still more clearly and impressively the action of mechanisms is seen in rhythmical fluctuations which are less characteristic for insects than for certain rodents, especially for species of the Leporidae and Microtinae. Owing to a fast succession of litters and to an early maturity, these species have a high fertility which causes a rapid increase in abundance. After it has reached a certain level, a conspicuous breakdown of the population takes place, the causes of which have been disclosed to a certain degree by the work of Green, Larsson, Chitty, Frank, and other authors. The population that has increased to high abundance begins to suffer from shortage of food which first causes a decrease in weight and draws on the reserves of the body; secondly, requires increased expense of energy to get food and room; and, lastly, intensifies the intrapopulation competition. For the individual, this means increased activity and psychic excitements which influence the endocrine system; by way of the hypophysis the adrenal gland rind is caused to produce more adrenalin which leads to reduction of the glycogene reserves, and finally to the lethal hypoglycemic shock (Frank, 1954, pp. 346–47). The inter-action of high fertility and of this physiological-pathological process, called "shock disease," may be considered as a classical example of the regulation of abundance by means of mechanisms.

Even if fertility and density-bound mortality interact so clearly, influences also become effective which are by no means dependent on the population and on the changes of its density. In certain regions of Germany the field vole, *Microtus arvalis* (Pallas), shows fluctuations in a cycle of three years which is based on the mechanism described. As Maercks (1954) proved, the cycle is disturbed by influences of the weather: unfavorable weather, especially in winter, retards or totally suppresses the maximum, whilst continually favorable weather hinders the breakdown of the gradation. Weather cannot be influenced by the population or its abundance; there is no causal connection from the popula-tion to the weather although a reverse connection does exist. Therefore an influence of the weather has to be considered in regard to the population as a random factor.

All observations in the field and particularly special investiga-tions show that chance interferes to a great extent with the development of every insect population. As one example — out of many which could be enumerated — the investigations of Subklew (1939) on the population dynamics of *Bupalus piniarius* (L.) in 1937 in the Letzlinger Heide may be cited. About two thirds of the deposited eggs were destroyed by *Trichogramma evanescens* (Westw.), which was only possible because the weather conditions were particularly favorable and because there was a stock of other host insects which from the beginning guaranteed a high abundance of the parasite. In a forest district severely infested by *Lymantria monacha* (L.), the strange case occurred that in consequence of the prodigal eating of the nunmoth caterpillars the needles, on which the geometrids had deposited their eggs, fell to the ground and the young caterpillars hatching there perished. Without exception the caterpillars suffered severe losses by violent thundershowers; also the pupae were parasitized in a high degree by *Ichneumon nigritarius* (Grav.), the abundance of which is dependent on the existence of other suitable host insects. In all cases mentioned the occurrence of influences caused by weather and other insects was not dependent on the *Bupalus* population, at least, not in the beginning when the first attacks of polyvoltine parasites occurred. These influences as regarded from the point of view of the population were due to chance.

In the examples just mentioned chance played a large part in the dynamics of abundance, whereas the so-called mechanisms certainly maintain balance for *Microtus arvalis* and probably for *Bupalus piniarius*. But investigations recently carried out on the dynamics of abundance of some forest insects furnish examples that the population density is governed to a great extent or almost exclusively by chance, and that mechanisms have little importance. Only two investigations may be mentioned which in a particularly detailed manner are concerned with the factors governing the abundance of the species in question.

The investigation of Heering (1956) was made on an outbreak of the buprestid, *Agrilus viridis* (L.), which from 1948 to 1951 in some regions of Germany caused severe damage in beech forests. The beetle deposits its eggs on the bark. The hatched larva enters into the bark and cuts meandering galleries into the phloem and the soft tissues of the xylem. The entry of the larva into the phloem and its establishment there depends wholly on the vitality of the tree: the healthy plant reacts by excretion of sap which often washes out the attacking larva, or by producing a wound callus in which the intensively growing tissues squeeze the larva to death. If the plant is less vigorous, particularly after long periods of drought, the defense reactions are weaker and the larva is able to succeed in entering the tree. Compared with the condition of the tree, that is to say its capacity for defense, all other factors causing mortality in the various stages of development, especially predators, parasites, and diseases, are subordinate. Of decisive importance for the abundance dynamics of *Agrilus viridis* is the supply of suitable host trees. In the scope of this supply, within the *Agrilus* population a more or less intense competition may arise; that is, a mechanism may become effective. But for our considerations it is decisive that the extent of the supply depends wholly on influences which can by no means by governed by the population or its abundance. Heering proved that the cause of the outbreak of *Agrilus viridis* in Southern Germany was an extraordinary succession of extreme dry and warm growing periods in the time from 1946 to 1952. The vitality of the beeches was strongly decreased and accordingly the supply of suitable brood material was increased and taken advantage of by the beetles. When — and this is very interesting as well as important for our considerations — in the autumn of 1952 rainfall set in and the beeches regained their former vitality, they not only became unsuitable for new infestation, but even trees already infested became able to drown the larvae boring in the phloem by the stronger sap flow, so that the *Agrilus* population suffered a catastrophic setback within a short time. Consequently it was the weather which, influencing the brood fitness of the host plant, indirectly determined the increase as well as the decrease in the abundance; that is to say, chance.

The other example is furnished by an intensive but not yet published investigation by Schütte on the population dynamics of *Tortrix viridana* (L.). The eggs of the tortricid, deposited on the twigs of the oak trees, hibernate. In spring when the young caterpillars hatch they enter the buds, but are enabled to do so only when the buds have reached a definite stage of development. If this stage of development does not exist, the caterpillars perish. Therefore the abundance of the population is highly dependent on the coincidence between the hatching of the young caterpillars and the development of the buds. Both of these events depend on the weather, but not in a wholly corresponding manner; therefore the two can diverge from year to year and the coincidence can be good or bad. Besides the weather, the individual quality of the tree has a decisive importance for the development of the buds. In a stand of oak trees individuals can be observed which regularly develop their buds early in the spring or late; these relative developments are fixed genetically and differ from each other in the extremes by more than two weeks.

In a stand containing equal proportions of oak trees of the different developmental types, a part of the tortricid population will find in all years the suitable bud stage, so that, as far as coincidence is concerned, a more or less level abundance can be maintained. In fact, in test stands of such a kind Schütte found a scarcely varying population density, the dynamics of which were caused almost exclusively by opposing effects of parasitic and predacious organisms. In comparison with this, in stands preponderantly composed of one type of oaks, e.g., of late developing trees, the abundance of *Tortrix viridana* in the main depends on the good or bad coincidence between the time of hatching and the stage of buds. In such stands the population density shows great fluctuations in correspondence with the conditions of coincidence which varies from year to year; biotic factors are only of little importance, perhaps because there is not sufficient time to react upon the rapidly varying abundance of their host animals. In any case it has to be stated that the population density of *Tortrix viridana* is largely and often almost exclusively dependent first on the composition of the oak stand, and second on the weather in spring, that is to say on chance.

These and further results led to doubts as to whether Nicholson's and other authors' opinions that a balance could only be maintained by mechanisms, by processes governed by the density of the population, can be correct. The supposition arose that chance, i.e., influences entirely independent of the population and its density, more than has been accepted until now, also has a great importance as a regulating factor. To prove this supposition, a model experiment was carried out. From the beginning it was clear that such an experiment could only vaguely imitate the real circumstances, but it was expected that it could help to give an answer to the question asked.

The factors influencing the population were represented by marbles. White marbles represented favorable factors increasing population, black marbles represented unfavorable factors decreasing population. The marbles were put into a pot and taken out at random.

To carry out the experiment, the following numerical suppositions were made. The initial density of the supposed population is 10. Every white marble doubles the respective density, every black marble halves it; in other words, the arithmetical factor that the density has to be multiplied by is 2 for the white and $\frac{1}{2}$ for the black marble. Since in the course of a long time the population density remains equal, that is to say promoting and checking factors compensate, the number of the white and black marbles used has to be equal, and 20 of each sort were put into the pot. Within a generation, 19 factors became effective; consequently 19 marbles were taken out at random and, after the arithmetical factor was determined from the portion of the white and black ones and was multiplied by the population density, they were put back into the pot, so that the original number of 40 marbles was always maintained.

The procedure may be illustrated by an example: The first 19 marbles contain 9 white and 10 black ones; 9 white and 9 black marbles compensate, one black marble remains; accordingly the initial population 10 is multiplied by $\frac{1}{2}$, the second generation begins with the density 5. The second time, 11 white and 8 black marbles are taken out; 3 remaining white marbles yield an arithmetical factor of 6; the population density 5 increases to 30. And so on.

This was carried out a hundred times, simulating a succession of 100 generations. The result is illustrated in Figure 1. Promoting and checking factors acting at random result during 57 generations in densities varying to and fro, showing also greater deviations to higher abundance, but always returning close to the initial density. In the fifty-eighth generation an enormous raising of the density begins, the numbers remain henceforth in general on a high level, even when a severe sinking takes place in the seventy-third to seventy-sixth generation. The highest number reached is 559,512, the initial density was 10. This is an increase which, if we consider the real case of a forest insect, e.g., *Panolis flammea*, lies beyond every possibility. The food supply, ever so large, but always limited, does not allow it.

Therefore it was supposed that, if an abundance of 10,000 is reached, the present food supply will no more be sufficient and

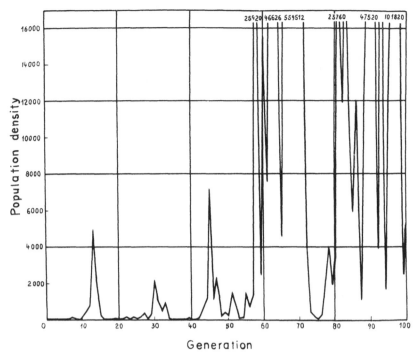

Generation

FIGURE 1: Fluctuations in numbers in a model experiment during 100 "generations." Chance factors only are involved. For explanation, see text.

the population density will decrease to the initial number 10 in consequence of deficiency of food through defoliation. Thus a mechanism is inserted into the play of chance: self-limiting of the abundance by food competition. Figure 2 illustrates the result of this change: up to the fifty-seventh generation the graph runs as in Figure 1. The increase in the fifty-eighth generation is limited at 10,000 — defoliation occurs, the population density decreases strongly because of deficiency of food — and the fifty-ninth generation begins again with the density 10. In the following generations the population density remains relatively low up to the 100th generation when the test was finished.

Anyone who is well acquainted with the numerical course of arhythmical fluctuations in the field will be impressed by the trend of the graph in Figure 2: it resembles exactly the graphs which were drawn for forest insects on the strength of annually

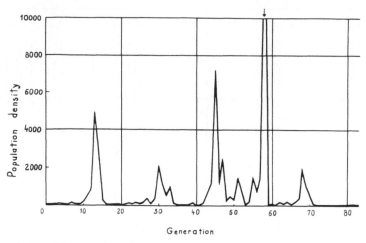

FIGURE 2: Fluctuations in numbers in a model experiment during 100 "generations." In addition to the play of chance factors, a controlling mechanism is introduced at the population level of 10,000. For explanation, see text.

repeated samplings. Figure 3 shows such graphs already published (Schwerdtfeger, 1941) for the 3 species, *Dendrolimus pini* (L.), *Panolis flammea* (Schiff.), and *Bupalus piniarius* (L.), which are important injurious insects in Germany. The graphs running through a period of sixty years or generations are based on samplings carried out every year in the winter in the same stands of the Letzlinger Heide, where caterpillars or pupae hibernating in the ground were found. The similarity of these graphs to those obtained by the test with the marbles cannot be overlooked and should lead to earnest thinking. If one could conclude from the resemblance of the phenomena a similarity of the causes, one would infer that the fluctuations of *Dendrolimus pini, Panolis flammea,* and other insects are affected almost exclusively by chance and that only occasionally, perhaps once in 100 generations, a simple mechanism represented by food deficiency is necessary to maintain the population density in balance.

The highly interesting result of the first model experiment caused further experiments of the same kind. To extend the questions asked, the presuppositions were varied. There is not enough space to present the results of these experiments in

detail; this must be reserved for a separate publication. Only
the more important results will be enumerated:

1. The other graphs obtained resembled, too, those shown in
Figures 2 and 3, provided that at a certain high level a reduction

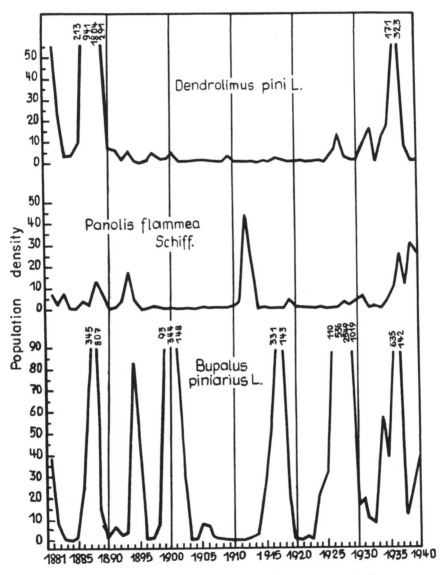

FIGURE 3: Population fluctuations of three forest insects in Germany,
1881 to 1940, from Schwerdtfeger (1941).

of the population density to the initial abundance as a result of food deficiency was supposed. Therefore, at high levels the occasional occurrence of a mechanism is necessary in order to prevent the abundance from increasing to an impossible level.

2. In contrast, at low levels no mechanism is necessary. By pure play of chance the initial population density 10 sank in the extreme to 0.000,000,3 and then rose in the course of 21 generations to 22.6; after a further 31 generations it had increased so far that the presupposed food deficiency mechanism had to become effective.

3. The larger the number of chance factors in operation, the rarer the necessity of a mechanism to reduce high abundance.

4. If instead of equal numbers, more promoting than checking factors were presupposed, mechanisms must become effective more frequently than if promoting and checking factors were equal.

Summarizing, it can be concluded from the series of model experiments, that a so-called balance in animal populations can be maintained during long periods by the mere effect of chance, particularly where the number of factors influencing the population is high. A mechanism is necessary only occasionally, as if it were a fuse, to prevent an excessive increase of the population.

The critic may object that this deduction goes too far and that an abstract model experiment, a test with black and white marbles, cannot yield a positive basis for it. The objection is correct, but it is permissible to say that the authors who regard only mechanisms as regulating processes base their opinion likewise mainly on abstraction, on experiments scarcely reproducing the real situation, and on play with figures. Starting points of this opinion are the arithmetical formulas concerning the interaction of host and parasite populations as well as laboratory investigations with appropriate conditions. A classical example in this regard is the recent publication of Nicholson (1954), which without doubt is highly rich in ideas and very stimulating; but its extensive deductions are based after all on some laboratory experiments with *Lucilia cuprina*. The results of such experiments and of arithmetical calculations are no doubt correct for the conditions presupposed. But frequently attention is not paid to the fact that these conditions are realized in nature only rarely or never. Nature works with far more complicated conditions than are represented by the rather simple presuppositions of the experimentalists and mathematicians. Accordingly the results of the experimenters and mathematicians are clear and regular, but the phenomena in nature are confusing and often do not follow any rule. One becomes skeptical when the patterns of population change considered to be characteristic and combined in Figure 6 of the recent publication by Nicholson (1954, p. 34) are in no case found to be realized in the fairly numerous analyses of population changes carried out in Germany on forest insects under natural conditions.

In nature, there are regularly numerous and varying factors influencing the population. They are partly dependent on the changing abundance, partly independent of it; they take effect on the population density partly in a promoting, partly in a checking manner. The greater the number of independent factors that influence the population, the higher is the possibility that promoting and checking factors compensate for some time, that the abundance remains in balance. Numerous observations in nature, of which only a few examples could be presented, and a series of simple model experiments permit the deduction that besides density-bound mechanisms, the so-called chance phenomena also have importance in the regulation of abundance, that the density of animal populations is governed by mechanisms as well as by chance, that the more multifarious the random play of chance becomes, the less important become the mechanisms.

## REFERENCES

Frank, F. 1954. Die Kausalitat der Nagetier-Zyklen im Lichte neuer populations-dynamischer Untersuchungen an deutschen Microtinen. *Zeitschr. f. Morph. u. Ökol. d. Tiere*, 43:321–356.

Heering, H. 1956. Zur Biologie, Ökologie und zum Massenwechsel des Buchenprachtkäfers *Agrilus viridis* (L.). *Zeitschr. angew. Entomologie*, 38:249–287; 39:76–114.

Maercks, H. 1954. Über den Einfluss der Witterung auf den Massenwechsel der Feldmaus *Microtus arvalis* (Pallas) in der Wesermarsch. *Nachrichtenblatt d. Deutsch. Pflanzenschutzdienstes*, 6:101–108.

Nicholson, A. J. 1933. The balance in animal populations. *J. Anim. Ecol.*, 2: Suppl. 132–178.

———. 1954. An outline of the dynamics of animal populations. *Austral. J. Zoology*, 2:9–65.

Schütte, F. Untersuchungen über die Populationsdynamik des Eichenwicklers *Tortrix viridana* (L.). *Zeitschr. ang. Entomologie*. (Im Druck).

Schwerdtfeger, F. 1941. Über die Ursachen des Massenwechsels der Insekten. *Zeitschr. ang. Entomologie*, 28:254–303.

Subklew, W. 1939. Untersuchungen über die Bevölkerungsbewerung des Kiefernspanners *Bupalus piniarius* (L.) *In* Schwerdtfeger, F., Der Kiefernspanner. 1937. Hannover, 1939, 10–51.

Thompson, W. R. 1956. The fundamental theory of natural and biological control. *Ann. Rev. Entomology*, 1:379–402.

# 3. Status of the Idea that Weather Can Control Insect Populations

## M. E. SOLOMON

The idea that weather controls populations of insects has always been widely held. Although entomologists generally are aware of the objections that have been raised to this view on theoretical grounds, some find it difficult to reconcile them with the obvious importance of weather as a cause of fluctuations and as a major influence upon abundance. In an attempt to clarify these matters, I shall discuss two questions: (1) To what extent must one's opinion that weather can control populations, or not, depend on one's definition of natural control? (2) What is the basis of the objection to the view that weather can control populations?

From *XI Internationaler Kongress für Entomologie, Verhandlungen,* 2 (August 25, 1960), 126–131.

CONSEQUENCES OF DIFFERENT IDEAS OF NATURAL CONTROL

If natural control is taken to mean that the abundance of an insect is observed to fluctuate, or that the numbers of a pest are sometimes reduced to a harmless level, then certainly weather factors, or any others that cause fluctuations in abundance, can be responsible for "control" in these senses. But usually the term implies more than mere variation in abundance, important though this is both to the ecologist and the economic entomologist. At the least, we may say with Thompson (1956, p. 379): "Natural control . . . refers to the fact that no organism increases indefinitely in number without limit." This leaves open the question whether the failure to increase indefinitely in numbers is the chance result of a continually changing pattern of environmental conditions, or whether density-dependent processes are essentially involved (i.e., the sort of adverse action that operates more severely per individual of the population as a result of raised density, with or without delay in this response to density change).

If it is believed that natural control has no essential element of density-dependent regulation, that it is simply the result of complex environmental forces that depress a population or stimulate its increase irrespective of what the level of abundance may happen to be at the time, then weather factors must be considered entirely competent to exert control, either alone or in conjunction with other nonregulatory factors. This is a viewpoint that no modern writer on the subject (so far as I know) adheres to without reservation. Thus Thompson (1956, p. 401), while holding that generally populations "are not truly regulated but merely vary," also allows that: "In extreme cases, where a chance conjunction of favorable circumstance has led to long continued increase, the induced shortage of requisites or the multiplication of natural enemies drawn in by the mass attraction of the host population, may reduce this population; but such cases are clearly exceptional." Andrewartha and Birch (1954, p. 660) make more allowance for the effects of shortage of resources such as food and nesting sites, but choose not to treat them as density-dependent processes; they adopt a similar attitude toward natural enemies. In this way they minimize the apparent significance of density-dependence. On the other hand, the fullest possible emphasis is given to the role of weather and other general environmental factors in natural control. They claim that some populations are not significantly influenced by density-dependent processes, but merely fluctuate in accordance with the weather and its influence on the environment. Birch (1958) has again expounded this point of view with reference to populations of a grasshopper. He maintains that, in these and in some other populations, abundance is limited entirely by the influence of weather on the environment, without the significant intervention of density-dependent processes.

If we adopt the more usual view, that natural control involves regulation by density-dependent processes, then it follows that weather alone cannot control a population. Weather alone could only act in a density-dependent way if the animals influenced the weather in some way. Of course, as many authors have pointed out, weather can act jointly with density-dependent processes in exercising control, e.g., by determining the extent of protective microhabitats and killing the surplus population that is excluded from them by the competition of other individuals. Correspondingly, in the system and terminology of Nicholson (1954, Figure 1), density regulation is jointly exercised by these two types of factors, which he calls density governing (density-dependent) and density legislative. The latter include factors like weather, not themselves influenced by population density.

Chitty (1960) maintains, with some reference to insects, that weather factors themselves act in a density-dependent manner. Briefly, the idea is that constitutional deterioration occurs as a result of high density, persists for some generations, and renders the population less viable in the face of any given set of weather conditions, and in the face of other factors such as pathogens. Hence the effect of weather depends partly upon the recent density of the population. How does this hypothesis differ from the ideas of Smith (1935), Nicholson, and others on the way in which weather factors and density effects jointly exercise natural control? In Nicholson's system, density effects are classified as density governing (density-dependent), the weather factors as legislative, the two acting jointly to impose natural regulation (natural control). Chitty holds that since the effects of weather can never be independent of the properties of the organisms, which are unlikely to be the same at all population densities, the action of weather factors is likely to be dependent on density. Personally, whatever practical difficulties may hinder the separation of density effects from those of weather in field studies, I consider the distinction on a theoretical level is justifiable and useful. But Chitty prefers to say that weather factors themselves act density-dependently. From this it is a short step to the claim (although Chitty does not make it) that weather can exercise natural control.

A more familiar basis for the statement that weather exercises natural control is that adopted by DeBach (1958, p. 476). He writes that "if intensity of unfavorable weather reduces the number of shelters and hence the number of insects surviving, weather acts in the manner of a density-dependent factor in that the higher the insect population becomes during favorable periods, the greater, percentagewise, will be the reduction during unfavorable periods. To sum up, weather may regulate insect populations by being of sufficient severity to restrict the size, quality and/or numbers of inhabitable spots in a given area." In a later passage, he mentions certain insects apparently being "regulated at low densities by weather interacting with micro-habitats." These statements omit mention of an essential third element, namely competition for the shelters, or dispersal from them as a result of crowding. This density-related element is an essential component of a realistic picture of the natural control of such insects. For if the population played an entirely passive role in the process, the weather would kill a proportion depending on the number and size of shelters available, and not on the general population density. Such action could not maintain natural control. It is clear from other passages that DeBach is well aware of this; he agrees (p. 475) that the objections to the idea of regulation by weather are correct "in the most technical sense," as opposed to a "general, more practical sense." But since "the most technical sense" seems to mean that in which clear definitions are implied, I suggest that any other sense will have to be avoided if confusion is to be overcome.

## OBJECTION TO THE HYPOTHESIS OF NONREGULATION

The state of a population simply varying in abundance according to changes in weather and other general factors, with no significant density-dependent regulation, may be called non-regulation. Andrewartha and Birch (1954) and Birch (1958) suggest that some populations remain in such a state and that their natural control can be accounted for without invoking density-dependence. Thompson (1956, p. 401) seems to hold a similar belief about populations in general, with the proviso already noted.

Consider an imaginary population free from density-dependent influences, and endowed with the high power of increase common to most species of insects. For a simplified analogy with natural diversity, consider the environment to be subdivisible into patches, each spatially uniform, but subject to the annual climatic cycle. On a patch where conditions are on the whole favorable, the animals will show an over-all trend of increase. From a patch where conditions are on the whole unfavorable, the animals will die out or move to the favorable patches. One can imagine that a minute proportion of a large number of patches might happen to provide conditions in which, over a number of generations, the successive net increases and decreases would balance, so that abundance fluctuated about a constant level. These conditions, or their mean over a period, would have to be matched very closely with the properties of the species concerned. Although such a state of affairs may be postulated in theoretical models like that of Schwerdtfeger (1958), it is scarcely conceivable that these precisely intermediate conditions would be widespread in the actual environment of any real population. Nor does the heterogeneity of natural environments, in space and time, offer an explanation of the stability of populations, relative to their great powers of increase. Why should the over-all resultant of environmental variations, however multifarious, add up over long periods to just the value required for approximate long-term stability of numbers, if none of them are geared to population size in a regulatory manner? Such a thing might be an expected result of natural selection, but only if each species were confined to environments of a particular degree of favorability, precisely matching its specific qualities; this is manifestly not what happens.

If there is a hidden fallacy in this seemingly axiomatic proposition, I have not been able to find it, nor, so far as I know, has anyone who read my earlier statement of these views (Solomon, 1957, p. 138). A more extended and systematic criticism of the various claims that have been made regarding the possibility of natural control without regulation has been published by Nicholson (1958a). The conclusion seems inevitable to me, as to many others, that density-dependent regulation must occur in all populations that persist for any considerable number of generations. And since weather factors are generally not themselves responsive to population density, they cannot exercise such regulation alone, but only in conjunction with density-dependent processes.

## A Proposal for Investigation

The rejection of nonregulation as an extremely improbable condition for any insect population to continue in for very many generations in succession does not force us to the opposite extreme. We are not compelled to conclude that populations are continuously being regulated at all times. On theoretical grounds we should suspect, as Huffaker (1958, p. 632) insists, that slight competition occurs even at very low densities. I do not wish to dispute this. But here I am concerned with effects that we may reasonably hope to detect in practice and that would be of some significance to an economic entomologist if he knew of their occurrence in a pest population. On this level, it seems to me probable that population regulation is often not only loose and variable, but also intermittent, and that the dynamics of populations in nature is often a more untidy and haphazard set of processes than Nicholson seems to allow. Some populations give at least a superficial semblance of unregulated variation over appreciable periods. But how are we to distinguish, in practice, between free fluctuation and fluctuation with a significant element of regulation? One answer is that the existence of regulation, its continuity or otherwise, and the resilience of the process, should all be demonstrable by means of field experiments in which certain populations are artificially increased or reduced (cf. Nicholson, 1958b, p. 326; Hairston, 1958, p. 327). A highly regulated population should soon return to its original density (or, if substantial environmental changes are occurring, to the density level of a comparable but untouched population). An unregulated population should retain the effects of an imposed alteration in density for a long time afterwards. Such experiments, carried out on a sufficiently representative variety of populations, should ultimately provide the required factual knowledge about the relative importance of regulation and unregulated fluctuation in various circumstances. Does it seem unrealistic to hope for considerable advances in this direction within, say, the next decade?

## REFERENCES

Andrewartha, H. G., and Birch, L. C. 1954. *The distribution and abundance of animals.* University of Chicago Press, Chicago.

Birch, L. C. 1958. *Cold Spring Harbor Symposia Quant. Biol.*, 22, 203–218.

Chitty, D. 1960. *Canad. J. Zool.*, 38, 99–113.

Debach, P. 1958. *J. Econ. Entom.*, 51, 474–484.

Hairston, N. G. 1958. In Reynoldson, T. B. 1958. *Cold Spring Harbor Symposia Quant. Biol.*, 22, 313–327.

Huffaker, C. B. 1958. *Proc. X Int. Congr. Entom.*, 2, 625–636.

Nicholson, A. J. 1954. *Australian J. Zool.*, 2 (1), 9–65.
————. 1958a. *Ann. Rev. Entom.*, 3, 107–136.
————. 1958b. In Reynoldson, T. B. 1958. *Cold Spring Harbor Symposia Quant. Biol.*, 22, 313–327.
Schwerdtfeger, F. 1958. *Proc. X Int. Congr. Entom.*, 4, 115–122.
Smith, H. S. 1935. *J. Econ. Entom.*, 28, 873–898.
Solomon, M. E. 1957. *Ann. Rev. Entom.*, 2, 121–142.
Thompson, W. R. 1956. *Ann. Rev. Entom.*, 1, 379–402.

# 4: Regulation of Animal Numbers: A Model Counter-Example

## HENRY S. HORN

The nature of mechanisms regulating the sizes of animal populations is a central problem in modern ecology. During the history of the formulation of this problem, workers and their ideas have become so polarized that new students are introduced to the problem as a "controversy." Students who lack the historical and philosophical perspective of their mentors are apt to judge the controversy as sterile, and consequently to regard the problem as insignificant.

The following model presents the regulation of animal numbers in such a way that the controversy becomes an aspect of the problem, rather than *vice versa*. The model shows in a generalized, but none the less analytical, way how the data of both sides of the controversy may be generated by plausible mechanisms that differ in degree, but not in kind. It also demonstrates that "density-

From *Ecology*, 49 (1968), 776–778.

dependence" and "density-independence" are not mutually exclusive alternatives. Finally, the flexibility of the model as a didactic tool is limited only by the imagination of the student and the diversity of nature. I am here concerned only with mechanism. The philosophy behind the positions taken by those engaged in the controversy is no less important, but it has been reviewed recently by Birch and Ehrlich (1967), Lack (1966), and Orians (1962). Techniques for analyzing censuses of animal populations have been presented by Salt (1966), Southwood (1967), and Tanner (1966).

This paper results from discussions with R. H. MacArthur. Critical comments by J. M. Emlen, D. J. and E. G. Horn, E. G. Leigh, and G. H. Orians were very helpful.

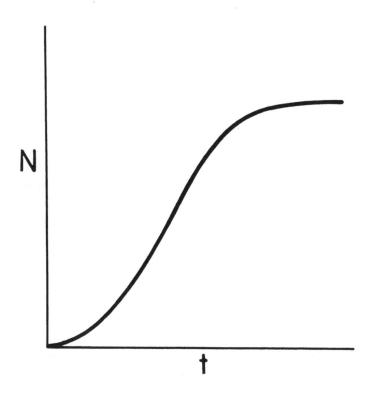

FIGURE 1

## THE MODEL

The "growth curve" of any animal population in a benign but finite environment will generally take the sigmoid form of Figure 1, where N is the size of the population and t is time. The well-known logistic is a special case of such a curve, but we need not be limited by stringent assumptions as to the formal shape of our curve; any biologically reasonable curve for which $d(\log N)/dt$ decreases monotonically with N will give the same qualitative result. We may now plot the slope of the sigmoid curve, $dN/dt$, against N for a benign environment, and use the resulting graph (Figure 2) to examine the stability of each population size. Where $dN/dt$ is positive N will increase, while where $dN/dt$ is negative N will decrease. N will converge on a stable population equilibrium, $N_e$, where $dN/dt = 0$. We shall call the curve of Figure 2 "g," as it represents a net growth rate in response to "density-dependent" factors. "Density-dependence" here means simply that the particular contribution to population growth rate is not a linear function of N.

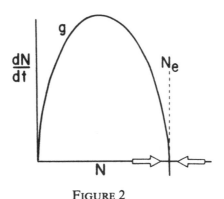

FIGURE 2

In a more rigorous environment, of course, the death rate will be increased at all population sizes by factors that are "density-independent." Strict "density-independence" implies a linear relationship between N and the particular component of $dN/dt$.

A density-independent factor, however, is somewhat more vaguely defined as a factor whose major component is linear. To present most forcefully the effect of a strictly density-independent factor on changes in population size, I shall assume that the effect of density-independent factors is indeed linear, but the model tolerates large departures from linearity before the qualitative results are appreciably changed. The death rate due to density independent factors is plotted as "di" in Figure 3. The population size at which the curves g and di intersect is now the population of stable equilibrium, $N_e$.

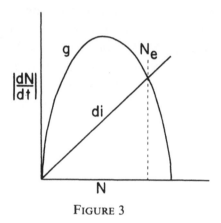

FIGURE 3

We may now represent fluctuations in density-dependent factors by movement of the peak and N-intercept of curve g. Similarly, fluctuations in density-independent factors may be represented by changes in the slope of curve di (see Figure 4). If g and di fluctuate independently and the average di intersects g near its peak (Figure 4a), the resulting fluctuations of the equilibrium population will follow the fluctuations of di more closely than those of g. If, on the other hand, the average di intersects g in its steeply descending region (Figure 4b), fluctuations of the equilibrium population will follow fluctuations of g more closely than those of di.

As an exercise to illustrate these results graphically, we may take relative positions of the g and di curves from a table of

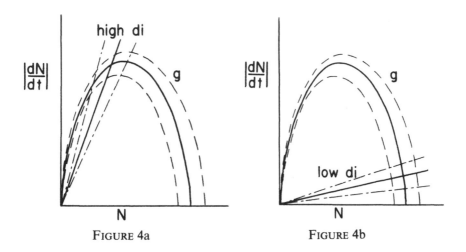

FIGURE 4a  FIGURE 4b

random digits. In Figure 5 we plot, as time series, the relative values of g and di, along with the resultant fluctuations in the equilibrium population taken from Figures 4a and 4b. It is clear from Figure 5 that fluctuations in the equilibrium population with a high sensitivity to the environment (high di) follow fluctuations in the density independent factors (di) more closely

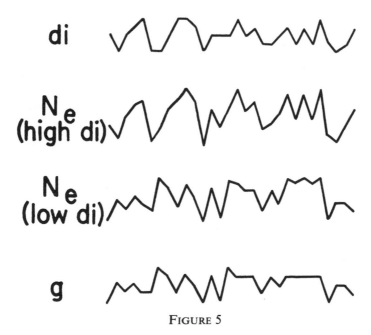

FIGURE 5

than they follow fluctuations in density dependent factors (g). Conversely, the population with a low sensitivity to the environment (low di) is closely tied to density-dependent factors (g).

## DISCUSSION

The vehemence of the controversy over the regulation of animal numbers suggests that there is some natural polarity in the mechanisms regulating the number of animals. Such a polarity may be generated by considering the effect of natural selection on the model. Natural selection is here defined to favor those individuals in each generation whose offspring form the greatest proportion of the breeders in the next generation.

If the density-independent sensitivity to the environment is high, the average equilibrium population is far below that which the biotic environment is capable of supporting, and populations are always fluctuating in response to a varying environment. Under these conditions selection favors those individuals with the highest intrinsic reproductive rate, even at the expense of a slightly greater sensitivity to the environment. As a result of such selection, the g and di curves should tend to become high and narrow if the di curve originally had a great slope. If, on the other hand, the density-independent sensitivity to the environment is low, the average equilibrium population is very close to the maximum number that the biotic environment can maintain. Selection will then favor the production of fewer, more vigorous offspring. Consequently, the g and di curves should tend to become flattened if the di curve originally had a low slope. Thus we might expect that animals would conform to the extreme models more often than to an intermediate model.

The suggested effect of selection is speculative and may be wrong, but the model itself allows some conclusions that are relatively independent of stringent assumptions about the form of the curves. The model shows analytically how "density-independent" factors may affect animal numbers in such a way that the best predictor of these numbers is a measurement of the value taken by the appropriate density-independent factor. But the model also suggests cautious interpretation of data that show, for example, a close correlation between some animal numbers and weather records. Given such data, we cannot conclude that density-dependent factors (e.g. predation and competition for space) are irrelevant to the regulation of numbers of these animals.

## REFERENCES

Birch, L. C., and P. R. Ehrlich. 1967. Evolutionary history and population biology. *Nature*, 214:349–352.

Lack, D. 1966. *Population studies of birds*. Oxford University Press, London. 341 pp.

Orians, G. H. 1962. Natural selection and ecological theory. *Amer. Naturalist*, 96:257–263.

Salt, G. W. 1966. An examination of logarithmic regression as a measure of population density response. *Ecology*, 47:1035–1039.

Southwood, T. R. E. 1967. The interpretation of population change. *J. Animal Ecol.*, 36:519–529.

Tanner, J. T. 1966. Effects of population density on growth rates of animal populations. *Ecology*, 47:733–745.

# 5.

# Effects of Population Density on Growth Rates of Animal Populations

## JAMES T. TANNER

Changes in size of animal populations usually follow seasonal and other variations in the environment. An important question is: Are these changes determined solely by the environment, or does the density of the population itself affect these changes? This study aimed at determining the relation between the rate at which a population grows or declines and the population density. Two approaches were used: (1) an examination of mathematical models of populations and (2) an analysis of the records of many different animal populations. In the second approach, data were obtained from the literature and statistically analyzed to test whether the growth rate of each population was or was not a function of population density.

From *Ecology*, 47 (1966), 734–740. By permission of the author, only parts of his paper have been reprinted. Extensive tables giving data, references, and results of statistical analyses have been left out, together with mention of these tables in the text. The second part of his paper, dealing with causes of population changes (which are extensively dealt with by other chapters in this book), is also omitted. The essence of his arguments and conclusions is preserved, and the reader is referred to Tanner's excellent original paper for documentation of his ideas.

The change in numbers of a population with respect to time, dN/dt, equals rN, where N is the number in the population and r is the rate of change per unit population; r = dN/Ndt. For conciseness r will hereafter be called a population's "growth rate," even though it can measure a decrease as well as an increase in population size. The growth rate equals the conventional birth rate (number born per unit population per time) minus the death rate (number died per unit population per time), and the growth rate will be positive, negative, or zero depending on the relative values of its two components.

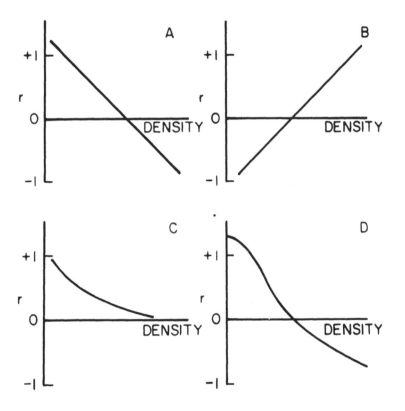

FIGURE 1: Graphs of r, population growth rate (r = dN/Ndt) as a function of population density. A. r is a decreasing linear function of density; B. r is an increasing linear function of density; C. r is a decreasing nonlinear function of density, found in growing Daphnia populations (Smith, 1963b); D. r is a decreasing nonlinear function of density, from the reproduction curves of Ricker (1954).

If r is a positive constant, the population will grow exponentially; if it is a negative constant, the population will decline exponentially; and if it is zero, the size will not change. Since populations never do any of these for an indefinitely long period, the value of r must vary from time to time. If its value is determined solely by the environment, r is independent of the population density. Otherwise r is some function of the population density. If r is a decreasing function of density, declining as the density increases as in Figure 1A, and if the environment is relatively stable for a sufficiently long time, the population will stabilize at the density where r is zero (or if there is a time lag, that is if r is determined by the density at some previous time, the population will oscillate about the central density). If r is an increasing function of density, as in Figure 1B, the populations will increase to infinity or decline to extinction, again assuming that r is not changed by environmental changes.

This research was completed while I was working in the Radiation Ecology Section, Health Physics Division, Oak Ridge National Laboratory; members of this section helped me to clarify many of the ideas expressed in this paper. Most of the statistical calculations were performed by the Mathematics Panel of the Oak Ridge National Laboratory.

## Rates in Mathematical Models

Mathematical models for animal populations present the population size, N, as a function of time, or the derivative of this with respect to time, dN/dt. The growth rate of the population, r, equals this derivative divided by N. The relation between r and N in these equations can be determined by differentiating r with respect to N. If the resulting derivative is negative, r is a decreasing function of N (Figure 1A); if it is positive, r is an increasing function (Figure 1B); if it is zero, r is independent of N. The only models examined here are those which have been found to describe the changes in actual populations.

The logistic equation, widely used in population studies, is discussed by Allee, Emerson, Park, Park, and Schmidt (1949, Ch. 21), who give numerous examples of this equation fitted to the growth of actual populations. The derivative, dr/dN, of the logistic equation is negative, showing that the growth rate decreases with increasing N, as in Figure 1A. Several persons (Gause and Witt, 1935; Smith, 1963b) have modified the logistic for one reason or another, but in these modified forms r remains a decreasing function of N.

Neyman, Park, and Scott (1958) constructed mathematical models for *Tribolium* populations based on extensive experiments in rearing these beetles; one model is for the numbers of a species living alone and another consists of two equations for two species of *Tribolium* living together. Nicholson and Bailey (1935) developed equations for the numbers of a host and its parasite; the first part of the oscillating curve predicted by these equations was followed by seven generations of a host-parasite population (DeBach and Smith, 1941). In all of these equations the derivative of the growth rate with respect to numbers is negative. Utida (1957) presented different equations which predicted the course of his oscillating host-parasite populations. The derivative for the parasite population is negative; that for the host population is positive or negative depending upon conditions other than the density of the host.

In summary, in mathematical models which have been shown to parallel the changes of actual populations, the derivative of r with respect to N is usually negative, i.e., the growth rate is a decreasing function of the density.

## RATES IN ANIMAL POPULATIONS

### Objective of Analysis

Few persons have appraised the direct effect of population density on population growth rates. Errington (1945, 1954) plotted per cent gain in Bobwhite and muskrat populations as a function of density and obtained curves showing smaller rates of increase with higher densities. Smith (1961, 1963a) found that in populations of thrips the growth rate over a month was negatively correlated with the population density at the start of the month. He also (1963a) showed that cultured *Daphnia* populations possessed growth rates that were nonlinear decreasing functions of density.

For this part of the study, I obtained from the literature the records of many different animal populations, calculated the growth rates from the recorded numbers, and tested the hypothesis that these rates were independent of the population densities.

## Methods of Analysis

A population record consists of a series of numbers, $N_0$, $N_1, \ldots . N_t, \ldots .$ At time t the population size is $N_t$ and its growth rate is $r_t$, which may be positive, negative, or zero. Expressing this rate in terms of the observed change in number, $\Delta N$, over the period between successive numbers $\Delta t$,

$$r_t = \frac{1}{N_t} \cdot \frac{\Delta N}{\Delta t} \tag{1}$$

Let $\Delta t = 1$, which can be done if the population was censused at regular intervals, then

$$r_t = \frac{N_{t+1} - N_t}{N_t} = \frac{N_{t+1}}{N_t} - 1. \tag{2}$$

Therefore $r_t$ is determined by the relation between $N_t$ and $N_{t+1}$. This relation needs to be examined to evaluate the effects of random variations in the data on the estimation of $r_t$.

$N_{t+1}$ must be some function of $N_t$ and the environmental conditions, which will vary with time. The nature of the function relating $N_{t+1}$ to $N_t$ and $E_t$, the environmental conditions affecting population growth at time t, is unknown, but to fit some biological facts the function has the following constraints: (1) if $N_t = 0$, $N_{t+1} = 0$; (2) an increase in $N_{t+1}$ over $N_t$ must have an upper limit which is a function of $N_t$, because the capacity for increase is limited; (3) a decrease cannot exceed $N_t$ in absolute value, because a population cannot go below zero.

A function of the following form meets the above conditions:

$$N_{t+1} = bN_t [f(N_t, E_t)]; b > 0, 0 \le f(N_t, E_t) \le 1. \quad (3)$$

The function in brackets, $f(N_t, E_t)$, denotes the interaction of numbers and environment. Because of random variation in the environment, variation independent of $N_t$, the successive values of this function will have some random variation. In the extreme case this entire function may vary independently of $N_t$; i.e., the coefficient of $N_t$ in $f(N_t, E_t)$ may be zero. Then for a given value of $N_t$, $f(0, E_t)$ will have an expected value denoted by $E(f)$ which is estimated by the mean value of $f(0, E_t)$. Under these conditions the expected value of $N_{t+1}$,

$$E(N_{t+1}) = bN_t [E(f)] = bN_t \overline{[f(0, E_t)]}. \quad (4)$$

Substitution of equation (3) in equation (2) shows that the relation between $r_t$ and population density is determined by $f(N_t, E_t)$, the interaction of numbers and environment. Substitution of equation (4) in equation (2) shows that, when there is no interaction between numbers and environment, $r_t$ will vary about its expected or mean value independent of population density.

Finally, it is possible for the successive values of N to be like a series of random numbers, with no correlation between $N_t$ and $N_{t+1}$. This may occur because of large and random errors in estimating population size or because the intervals between counts were so long that numerous intervening events destroyed any correlation between successive counts. If the successive values of N are like a series of random numbers, $r_t$ as determined by equation (2) will be a decreasing function of $N_t$. The proof of this is presented by Watt (1964). Therefore, to avoid obtaining fallacious results from a random series, it is necessary to check each population record to insure that the successive values are correlated and therefore that $N_{t+1}$ is some function of $N_t$. Each series of population counts examined in this study was tested for nonrandomness by the method of runs. Those in which the probability of randomness was not less than 5 per cent were eliminated.

The preceding arguments lead to the conclusion that a record of a population, $N_0$, $N_1$, ... $N_t$, ..., can be used to calculate the population's rate of growth for each interval, by equation (2), and that the relation between growth rate and population density may be determined statistically if the following three conditions are met: (1) $N_t$ and $N_{t+1}$ are correlated; (2) the population was censused at constant intervals so $\Delta t$ can have the value 1; and (3) the counts, censuses, or estimates reflect the actual density of the population.

From the population record will be obtained a series of $r_t$ and $N_t$, of growth rates for each interval and of numbers or density at the beginning of that interval. It would be logical to consider $r_t$ as a function of $N_t$ and to use regression methods to test the nature of this function. Regression methods, however, assume that the independent variable ($N_t$ in this case) is known or measured without error, and this certainly is not true for many population records. By contrast, calculation of a correlation coefficient does not require that one variable be free of error, and therefore correlation is preferable to regression in this analysis.

The correlation coefficient is applicable only when the two variables are linearly related or approximately so. Smith (1963b) found that in *Daphnia* populations r is the nonlinear function of density shown in Figure 1C. Ricker (1954) postulated a series of reproduction curves for different animals; from the curves which he considered to best fit the data from actual populations can be constructed the relation between r and density shown in Figure 1D. Both of these curves can be satisfactorily approximated by straight lines. After the computations had been completed, I read the paper by Morris (1963) in which he stated that, at least for the insect populations he studied, $\log N_{t+1}$ is a linear function of $\log N_t$. If this is so, then $\log (r_t + 1)$ is a linear function of $\log N_t$, and a correlation coefficient of these logarithms would be larger than one using actual values. To test this, seven population records, whose correlation coefficients based on actual values were not quite large enough to be significantly different from zero, were analyzed using logarithms. The resulting correlation coefficients were larger by an average of 13 per cent, but in only one case, that of the Fulmar, was the improvement great enough to produce a coefficient significantly different from zero. The correlation coefficients for these seven populations were calculated using logarithms. The above considerations justify the use of the correlation coefficient for testing the relation between growth rate and population density.

If two or more population records of the same species are available, the best estimate of the correlation can be obtained by calculating a joint correlation coefficient, as follows. First, since the growth rate is expected to be related to the density of each population, each $N_t$ was divided by the average size of the population; dividing by the area inhabited would have been better but this was usually unknown, and the area inhabited should be directly proportional to the average population size. Second, the sums of cross-products and of squares used in calculating the joint correlation coefficient involved the deviations from the separate means of each population.

In this study, a correlation coefficient was calculated for the paired variables, $r_t$ and $N_t$, of each population record (or a joint correlation coefficient for two or more records of the same species). The null hypothesis that the correlation coefficient is zero was tested. If the null hypothesis was accepted, $r_t$ was considered to be independent of $N_t$. If the null hypothesis was rejected, $r_t$ was identified as a decreasing function of $N_t$ if the correlation coefficient was negative, or an increasing function if the coefficient was positive.

## Sources of Data

The data were obtained from censuses of animal populations that met the following requirements: (1) actual counts of a population inhabiting a definite space so that density was directly proportional to the number, or a reliable index of the density of the population; (2) counts or estimates made periodically so that the growth rates were based on constant intervals ($\Delta t$); (3) a sufficiently long series of such counts to enable a meaningful test despite random variations.

About half of the populations analyzed were censused. For

the remainder an index of density was available. This index for the fish populations was calculated from the annual commercial catch corrected for variations in fishing intensity. Fur returns were used as indexes of density for some northern mammals; Keith (1962) gave reasons for believing that these were valid indexes of density when the returns came from constant geographic areas where fur prices had not failed. Hunting kill statistics were not considered to be a reliable index of density because of variations in hunting pressure with periods of war, changing economic conditions, and changing recreational habits. Exceptions to this were some bag records made on British game preserves which had been managed intensively for decades; Keith (1962) believed that these were reliable indexes of density. Fur returns from the Hudson Bay Company and game bag records for some British preserves supplied the longest series of records for this study.

I found records of 111 populations, representing 71 species, that met the requirements of this study.

## Results and Discussion of Results

Most results were obtained by using each entire population record without regard to changes in the environment. If different periods of a population's existence possessed distinctly different environmental conditions known to affect reproduction and survival, a better estimate of the correlation between growth rate and density can be made by a joint correlation coefficient, in which the deviations are measured from the means for the different conditions. This was done for two species, thrips and Starling. The number of thrips (Davidson and Andrewartha, 1948) increased each year to a high level during the Australian spring and then dropped to a low level during approximately 10 months of drought and low temperatures. The Starling population inhabited an area where the death of many trees beginning about the sixteenth year of the series (Kendeigh, 1956) produced many more nesting sites and a period of relative abundance of these birds. The joint correlation coefficients for these two species are very significantly different from zero although the correlation coefficients calculated without regard to periods of abundance and scarcity are not so. This procedure can be justified only when there is direct evidence of environmental changes.

Another situation requiring separate analysis is that of five species of insects. Populations of these five species are similar in that they occasionally erupt, becoming very abundant for short periods, after which the numbers fall to low levels for comparatively longer periods. The calculations using the entire population record irrespective of population level, indicate a growth rate independent of density. The results are different, however, when the population record of each of these species was divided into periods of scarcity and abundance, and the correlation analysis was conducted only on the data from the long periods of scarcity. The results are that four of the five correlation coefficients from periods of scarcity are negative and significantly different from zero.

The results are summarized in Table 1. Of the 64 species remaining after 7 were eliminated because their population records were not different from a random series, 47 species had coefficients that were significantly different from zero and were negative. Only one, the human population of the world, which has been increasing at a greater rate with the increase in popula-

TABLE 1: *Summary of Results* [a]

| Category | A | B | C | D | E |
|---|---|---|---|---|---|
| Invertebrates other than insects | 5 | 1 | — | — | 4 |
| Insects | 17 | 2 | 3 | — | 12 |
| Fish | 7 | — | 3 | — | 4 |
| Birds | 23 | 4 | 5 | — | 14 |
| Mammals | 19 | — | 5 | 1[b] | 13 |
| Totals | 71 | 7 | 16 | 1 | 47 |

Column A: Total number of species; B: Number of species eliminated because the population record was not significantly different from a random series; C: Correlation coefficient not significantly different from zero, probability of the null hypothesis exceeding 0.05; D: Coefficient positive and significantly different from zero; E: Coefficient negative and significantly different from zero.

[a] Table 3 in original.
[b] The human population of the world.

tion density, has a coefficient that is significantly positive. Furthermore, of the 16 species for which the correlation coefficient did not differ significantly from zero, only one did not have a negative estimated value. Population rates of change are usually, therefore, a decreasing function of population density; as can be seen from Table 1, this conclusion applies to animals which are taxonomically very different.

## REFERENCES

Allee, W. C., A. E. Emerson, O. Park, T. Park, and K. P. Schmidt. 1949. *Principles of animal ecology*. Saunders, Philadelphia. 837 pp.

Davidson, J., and H. G. Andrewartha. 1948. Annual trends in a natural population of *Thrips imaginis* (Thysanoptera). *J. Animal Ecol.*, 17:193–199.

DeBach, P., and H. S. Smith. 1941. Are population oscillations inherent in the host parasite relation? *Ecology*, 22:363–369.

Errington, P. L. 1945. Some contributions of a fifteen-year local study of the Northern Bobwhite to a knowledge of population phenomena. *Ecol. Monogr.*, 15:1–34.

———. 1954. On the hazards of overemphasizing numerical fluctuations in studies of "cyclic" phenomena in muskrat populations. *J. Wildl. Mgmt.*, 18:66–90.

Gause, G. F., and A. A. Witt. 1935. Behavior of mixed populations and the problem of natural selection. *Amer. Naturalist*, 69:596–609.

Keith, L. B. 1962. *Wildlife's ten-year cycle*. University of Wisconsin Press, Madison. 201 pp.

Kendeigh, S. C. 1956. Census 36. *Audubon Field Notes*, 10.

Morris, R. F. 1963. Predictive population equations based on key factors. *Mem. Entomol. Soc. Canada*, 36:16-21.

Neyman, J., T. Park, and E. L. Scott. 1958. Struggle for existence; the Tribolium model: biological and statistical aspects. *Gen. Systems*, 3:152–179.

Nicholson, A. J., and V. A. Bailey. 1935. The balance of animal populations. Part I. *Proc. Zool. Soc. London*, 1935:551–598.

Ricker, W. E. 1954. Stock and recruitment. *J. Fish. Res. Bd. Canada*, 11:559–623.

Smith, F. E. 1961. Density dependence in the Australian thrips. *Ecology*, 42:403–407.

———. 1963a. Density dependence. *Ecology*, 44:220.

———. 1963b. Population dynamics in *Daphnia magna* and a new model for population growth. *Ecology*, 44:651–663.

Utida, S. 1957. Population fluctuation, an experimental and theoretical approach. *Cold Spring Harbor Symposia Quantitative Biol.*, 22:139–151.

Watt, K. E. F. 1964. Density dependence in population fluctuations. *Can. Entomol.*, 96:1147–1148.

# 6. Endocrines, Behavior, and Population

## JOHN J. CHRISTIAN
## DAVID E. DAVIS

For several decades the spectacular increase and decrease of certain arctic mammals has stimulated research on populations. The crashes of rabbits were dramatized by Seton (1), and the suicidal movements of lemmings were publicized by many authors. However, as is so often the case, the conspicuous features turn out to be merely an extreme case of a very general phenomenon — namely, the fluctuations of a population. Investigators first sought an explanation for the "crash," but now most of them search for a description and understanding of the interaction and relative importance of the many factors that influence the ups and downs of populations.

From *Science*, 146 (December 18, 1964), 1550–1560. Copyright 1964 by the American Association for the Advancement of Science. With the permission of the authors, the first six figures of the original paper have been omitted, along with mention of them in the text. The remaining figures are renumbered accordingly. The omitted figures depict histological structure of mouse adrenals, and the reader may verify certain statements in the text by referring to these figures in the original paper.

In this chapter we describe the current status of our understanding of population fluctuations, emphasizing the regulatory features that prevent populations from destroying the habitat. The research discussed is limited to work with mammals, since the mechanisms are best known for that class. It is assumed that the reader has knowledge of ecological principles such as density-dependence and limiting factors.

For many years it was assumed that epizootics, famine, and climatic factors terminated the explosive rises in population size and precipitated the often spectacular crashes (2). However, by the early 1940s it had become apparent that none of these mechanisms explained some of the observed declines in population, and it was suggested that factors intrinsic to the population were involved in its regulation (3). The skepticism toward earlier explanations was reflected further in a review by Clarke in 1949 (4), as well as in Elton's classic earlier work (5). Probably the greatest shift in emphasis has occurred since 1949; there has been an upsurge of investigations in which density-dependent changes in the animals themselves have been explored, and of theories in which the observed phenomena of population growth and decline (6–9) are explained in terms of biological mechanisms intrinsic in the populations and not only as results of the action of external factors. It is clear that food, climatic factors, and disease may cause population change. Indeed, it would be foolish to state that these factors do not, under certain circumstances, limit population growth or produce spectacular decline. The early investigations of Emlen, Davis, and their co-workers (8) on populations of Norway rats demonstrated clearly that environmental factors can reduce a population. For example, a drought followed by excessive rain resulted in a notable decline in rats in Baltimore (8). However, as early as 1946 spectacular declines in rat populations were found to be coincident with social disturbances rather than with environmental changes.

The suspicion that social phenomena were involved prompted a search for mechanisms that could regulate the growth of populations in a density-dependent manner. No longer is attention focused exclusively on spectacular crashes and the causes of death. Instead, an attempt is made to integrate the social actions and the well-known habitat factors into a scheme that will explain the changes in populations. Since social or behavioral features are density-dependent, they become evident only at high population levels. Nevertheless, such features are present in low populations, but inconspicuous. Purely ecological factors, such as food and climatic conditions, also affect populations and, indeed, may prevent a population from attaining a level where social forces can become important. Hence, examination was begun of a theory which states that, within broad limits set by the environment, density-dependent mechanisms have evolved within the animals themselves to regulate population growth and curtail it short of the point of suicidal destruction of the environment (6, 10–13). Milne (12) has summarized this point of view as follows: "The *ultimate* capacity of a place for a species is the maximum number of individuals that the place could carry without being rendered totally uninhabitable by utter exhaustion or destruction of resources...." The environmental capacity cannot be greater than ultimate capacity; it could conceivably be equal to ultimate capacity but ... is usually somewhat smaller." We would modify the "somewhat" to "considerably," in view of the situation most often observed for mammals (here we are talking primarily of herbivores and rodents). Milne goes on to say that "the one and only perfectly density-dependent factor [is] intraspecific competition."

While some investigators ascribe all regulation and limitation of populations to direct effects of environmental factors, others recognize that a feedback control of population growth exists. However, there is not complete agreement on the mechanisms by which these results are achieved. In the rest of this article we review the more recent results of experiments made to test the hypothesis that a behavioral-physiological mechanism operates to control population growth in mammals, and we consider criticisms of this view in the light of the evidence on which they are based. The acceptability of the hypothesis should be considered from the viewpoint of what would constitute disproof. To prove that behavioral mechanisms *never* affect population growth is of course impossible. To cite one or more cases in which some habitat factor controlled the population is merely an elaboration of the obvious. Thus, proof or disproof of the hypothesis reduces to the problem of finding how frequently and under what circumstances the behavioral mechanism does operate. The discovery of other physiological mechanisms [for example, pregnancy block caused by the proximity of strange males (14) or direct block of reproduction organs in *Peromyscus* (15)] does not alter the situation. Similarly, the absence of the mechanism in certain mammals would not prove its absence in rodents. The problem, then, is not that of proving the existence of a behavioral-physiological mechanism but that of proving the importance of such a mechanism in the regulation of populations.

## PHYSIOLOGICAL MECHANISMS

On the basis of the knowledge of pituitary-adrenocortical physiology available prior to 1950, it was proposed (16) that stimulation of pituitary-adrenocortical activity and inhibition of reproductive function would occur with increased population density. It was suggested, further, that increased adrenocortical secretion would increase mortality indirectly through lowering the resistance to disease, through parasitism or adverse environmental conditions, or, more directly, through "shock disease," although it soon became evident that unwarranted emphasis was being placed on "shock disease." Implicit in this theory and in the design of experiments to test it was the theory that behavioral factors (aggressive competition, for example) comprised the only stimulus to the endocrine responses which would invariably be present in every population. Experiments to test the theory were conducted on animals which were provided with (or known to have) more food, cover, and other environmental assets than they could utilize, and were thus in populations either really free of predation or having a minimum degree of predation (17–19).

The endocrine responses were first assessed through measurement of changes in the weights of the adrenals, the thymus, the reproductive organs, and certain other organs. Interpretations of adrenal weights are reliable and simple in species that have been adequately studied in the laboratory — for example, in rats and mice, whose adrenal physiology and morphology have been examined in detail under a variety of circumstances. In particular, the immature zonation (X-zone) of mice and its changes with respect to age and sex had been thoroughly explored (20). An important point was the lack of evidence of function for this zone. Where adrenal weight could be reliably interpreted in terms of function, it seemed better, in the study of populations, to use an indicator of long-term conditions, rather than indicators highly sensitive to acute stimuli. For example, concentrations of ascorbic acid in the adrenal gland and concentrations of corticosteroid in plasma respond very rapidly to acute stimuli. Furthermore, the interpretation of changes in adrenal weight was supported by other morphological criteria of increased corticosteroid secretion, such as involution of the thymus, though the possible role of other factors in the alteration of these other organs was not overlooked. Nevertheless, even in rats and mice, changes in adrenal weight can only be considered strong presumptive evidence of changes in adrenocortical function until validation is obtained by direct functional studies.

Adrenal weights are not valid indices of function unless certain precautions are observed. The presence of immature zones (X-zones) complicates the use of adrenal weights as indices of function, since evidence that such zones contribute to cortical function is lacking. Another complication is the possibility of weight loss with sudden or excessive stimulation. Moreover, there may be a misleading increase in adrenal weight due to accumulation of lipids with cessation of adrenocorticotropic hormone (ACTH) stimulation. Also misleading is the hypertrophy of the adrenal medulla which occurs in some instances, but this usually is not important (21). In addition, qualitative changes in the corticosteroids secreted may require modification of interpretations based on adrenal weight. Finally, sexual maturation or activity may alter cortical function and adrenal weight. Androgens involute the X-zone or decrease adrenal weight in adult animals, whereas estrogens commonly increase adrenal weight. It is axiomatic that, in comparing changes in adrenal weight with changes in population, one must consider adrenal changes due to reproductive condition, and that only adrenals from animals of similar reproductive status can properly be compared.

In addition to these physiological considerations, there is the problem of obtaining adequate samples. Since there are two sexes and at least two age groups, the sample must contain enough animals in each of four categories for appropriate analysis. This requirement may seem obvious, but it often has been neglected.

The foregoing principles regarding the interpretation of adrenal weights have been presented because in many studies one or more of these principles has been neglected. Earlier work on physiological responses to changes in populations has been reviewed elsewhere (6, 17–19) and is only summarized here. In experiments with mice in the laboratory, progressive adrenocortical hypertrophy and thymic involution were observed to occur with increasing size of population. Somatic growth was suppressed and reproductive function was curtailed in both sexes. Sexual maturation was delayed or, at higher population densities, totally inhibited. Spermatogenesis was delayed, and the weights of the accessory sex organs declined with increasing population density. In mature females, estrous cycles were prolonged and ovulation and implantation were diminished; intrauterine mortality of the fetuses increased. Recent results in rabbits show an increase in intrauterine mortality in association with increased population density, especially in the fetuses of socially subordinate females (22). In another study a similar increase in intrauterine mortality was noted, but no difference in rate of resorption of embryos relative to social rank was observed (23). Increased resorption of embryos also followed grouping of *Peromyscus* (24). However, in mice, the importance of resorption of embryos in regulating birth rates may vary considerably from population to population (17). Also, increased population density resulted in inadequate lactation in mice, so that nurslings were stunted at weaning. This effect was seen again, though to a lesser degree, in animals of the next generation not subjected to additional crowding (25). It has since been found that crowding of female mice prior to pregnancy results in permanent behavioral disturbances in subsequently conceived young (26). Particularly interesting in this regard is the observation that increased concentrations of corticosterone may permanently affect the development of the brain in mice (27). Increased population size also delayed or totally inhibited maturation in females, as well as in males, so that in some populations no females reached normal sexual maturity. The combination of these responses, believed to result from inhibition of gonadotrophin secretion, resulted in a decrease in birth rate, or an increase in infant mortality, or both, as populations increased, until increase of the population through the production of young ceased. Concentrations of gonadotrophins in relation to changes in population size have not been measured. However, increase in the number of rats per cage was found to alter responses to injected gonadotrophins, even when the area per rat was kept constant (28).

Increased population density may affect reproductive function in male and female house mice differently in different populations. The growth of one population was slowed and eventually stopped mainly by a decline in birth rate due to (i) failure of the young to mature and (ii) decrease in the reproductivity of mature animals (17). Infant mortality was a negligible factor in this population. In several others a decline in the survival of nurslings was largely responsible for a slowing and stopping of population growth, although a lowering of the birth rate also occurred (17, 19). In most populations both a decrease in birth rate and a decrease in the survival of nurslings contributed importantly to slowing of the rate of population growth and limitation of numbers, but, as one might expect, the relative importance of these two factors varied among populations. In populations in which a change in birth rate was the main regulating factor, other measurements indicated that it was the males which were primarily affected by increased population density, the effect on females being slight. When increasing mortality of nurslings was the main regulating factor, the females were severely affected and the males were relatively less affected than in other populations (17). These results imply that effects on the male may be important in producing declining birth rates, although failure of females to mature also would contribute to a decline in birth rate in any population and cannot be excluded. Final conclusions regarding this problem must await further investigation.

For many years it has been known that disease sometimes becomes rampant when populations reach peak levels (5). However, the belief that disease usually is a primary cause in the reduction of populations has not been supported (5, 11). A change in host resistance has been suggested as an underlying condition leading to increased mortality from epizootics (6, 11, 17). It is well known that glucocorticoids reduce resistance to infectious disease by inhibiting the normal defense reactions. They may also be involved to some extent in the pathogenesis of other disease, such as glomerulonephritis as seen in woodchucks (17). Furthermore, grouping, presumably through adrenal stimulation, augments adrenal-regeneration hypertension in rats (29). Experiments have shown that, with increased population density, there is a marked depression of inflammatory responses, of formation of antibodies, and of other related defenses, with a resultant increase in susceptibility to infection or parasitism. For example, in a confined population of rabbits a highly lethal epidemic of myxomatosis occurred coincident with attainment of a high density (22). During this epidemic dominant animals and their descendants had the highest survival rate, implying a breakdown in host resistance following increased social competition. Similar results were observed in a population of deer, associated with high densities and subsequent decline in population (30). Increased density also enhances mortality from other causes — for example, radiation, amphetamine toxicity, and toxicity due to other pharmacologic agents (31). Decreased resistance to amphetamine following grouping is probably due to increased secretion of epinephrine and not to increased secretion of corticosteroids (32). Emotional stress also enhances mortality from disease, probably through the same endocrine mechanisms (33). These results suggest that at high population densities an epidemic occurs in part because resistance is lowered. Thus, disease is a consequence of high population rather than a primary cause of a decline in population.

## BEHAVIORAL ASPECTS

What basic behavioral factors result in these profound effects? It seemed to us that any density-dependent effects would be related to social rank. Experiments made to test this hypothesis showed that adrenal weight and somatic growth were related to social rank (18, 34). Other experiments, in which adrenocortical function was assessed from counts of circulatory eosinophils (35), confirmed these results. Adrenal cortical activity is similarly related to social rank in rats and dogs, as determined by lipid and cholesterol concentrations in the adrenals of rats and by hydrocortisone secretion in dogs (36). In several somewhat related experiments it has been shown that the degree of response to changes in population size is dependent on the behavioral aggressiveness of the strain or species involved (19, 37). In the highly aggressive house mouse (*Mus musculus*), changes in adrenal weight, ascorbic acid content, and cholesterol content demonstrated the important role of behavioral factors in the responses to changes in population density. In contrast, deer mice (*Peromyscus maniculatus bairdii*) failed to respond, due to behavioral characteristics and not to an inherently unresponsive endocrine system (37). The two species responded equally when exposed to trained fighters of their own species or when subjected to cold.

In most studies of social rank an indirect measure of adreno-cortical function was used, such as the weights of adrenal and thymus, cholesterol and ascorbic acid content of the adrenal, and numbers of circulating eosinophils. Recently, a number of investigators have observed increases in adrenocortical function with increases in population density. There is an appreciably greater in vitro production of corticosteroids by adrenals in grouped mice than in singly caged mice (38). Albino laboratory rats show an increase in plasma corticosterone concentrations from 6.7 to 22 micrograms per 100 milliliters when they are maintained in colonies rather than in groups of four to a cage (39). There was also a fivefold increase in the in vitro production of corticosteroids by the adrenals of the colony-maintained rats. Barrett and Stockham (40) reported a 73 per cent increase in plasma corticosterone concentrations, as measured fluoro- metrically, in albino rats kept in groups of 20 as compared with concentrations in single caged animals. Pearson (41) found that, in general, plasma corticosterone levels increased with increasing density in freely growing populations of mice, although there was considerable scatter in the results, possibly because of capture and handling procedures. Thus, direct measurement of corticosteroid levels confirms conclusions from experiments in which morphological criteria were used to assess adreno- cortical function in Norway rats and house mice.

Increases in the weight of the spleen in response to increased population density have been reported in mice and voles (6, 42, 43). In house mice the increase in splenic weight is due to increased hematopoiesis involving all blood-forming elements, and not solely to erthyropoiesis, as in voles (43). The increase probably is related to social rank (44), although a response to injuries from fighting could not be ruled out.

The problem of the role of food invariably arises in discussion of changes in population. A shortage of food might have the direct effect of causing starvation or an indirect effect by increas- ing competition among animals. Contrary to a widely held belief, chronic inanition per se (as opposed to acute starvation) appears not to result in increased adrenal weight or increased cortical function in rats, mice, and men (45, 46). Experiments with mice showed that chronic inanition had no effect on adrenal weight, either directly or indirectly (46). However, inanition curtailed reproductive function independently of its effects on the pituitary-adrenocortical system. In some species, limitation of the food supply apparently increases competition (22, 47, 48), and thus subordinate animals are more affected by the shortage than dominant ones. Resistance to starvation (and thus survival) is greater in dominant or older animals than in subordinate or younger animals (22). Also, the decreased need for protein seen in deer during winter and early spring is frequently overlooked (49). It is possible that some microtines or other rodents also have mechanisms for taking advantage of bacterial protein synthesis during periods when proteins and natural plant foods are scarce. On the basis of existing evidence (11, 50), the direct effect of food shortages cannot be considered a common denominator in the regulation and limitation of growth of populations of herbivorous mammals. Studies of populations of *Clethrionomys* (51, 52), lemmings (53), voles (11, 19), woodchucks (54), *Apodemus* (52), and other mammals have shown that a deficiency of food either was not a factor in population decrease or else had an effect complementary to behavioral changes associated with changes in population density (47). From evidence currently available it appears that the effects of restricting water intake over a long period can be regarded in the same fashion as the effects of chronic inanition. In a thorough study of food requirements and availability of food in relation to populations of small mammals, it was shown that food was not a limiting factor in the area studied (50). More critical studies of this sort are needed before a final evaluation can be made of the relative importance of food shortages in limiting population growth and of the degree to which such limitation, when it does occur, is associated with increasing competition within existing hierarchical structures.

The important point, in assessing the effects of behavioral factors on adrenal function, is the number of interactions between individuals rather than density of population per se. Thus, age, sex, previous experience, local distribution, and other factors may be critical in producing effects (6, 17–19, 55). The development of the adrenal responses may be produced by very brief encounters with other animals. Experiments showed that 1-minute exposure to trained fighter mice, 1, 2, 4, and 8 times a day for 7 days produced increases in adrenal weight and increases in adrenal and plasma concentrations of corticosterone (56). As few as two 1-minute exposures per day resulted in a 14 per cent increase in adrenal weight, and eight exposures daily resulted in a 29 per cent increase. Plasma corticosterone increased by 67 per cent. Adrenal levels of corticosterone increased in proportion to adrenal weight, so corticosterone concentrations per gram of adrenal tissue remained constant. These results validate, for mature male house mice, the use of adrenal weight as an index of cortical function. Thus, a few short daily exposures to aggressive mice produced a greater increase in adrenal weight than caging male mice of the same strain together in groups of eight continuously for a week (37, 56). These results should serve as an answer to the criticism that laboratory experiments on populations are not realistic because of artificially high densities.

Differences in basic aggressiveness of the strain or species must be considered in a comparison of relative population densities. For example, albino mice are extremely docile in contrast to some strains maintained in the laboratory (57). *Peromyscus maniculatus bairdii* also is nonaggressive — even more so than albino *Mus* (37). Recently Southwick has demonstrated the importance of behavioral factors in eliciting an adrenocortical response in *P. leucopus* by showing that grouped animals had no adrenal response when they were "compatible" but did if they were "incompatible" (58). Thus, to compare absolute densities in the laboratory with those of feral populations is not a justifiable procedure.

It is often said that fighting per se, or injury from fighting, produces the endocrine changes that occur with change in rank and number (59). However, data from a large number of populations of mice demonstrate that the endocrine responses to grouping are identical whether or not there is fighting or injury (6). Fighting is another symptom of social competition. It seems clear that the basic stimulus to the endocrine changes are sociopsychological, or "emotional," and not physical in nature. Pearson has made the interesting observation that in freely growing populations a few excessively submissive, thoroughly beaten-up, badly scarred mice have low plasma concentrations of corticosterone (41). This result agrees with our observations that mice that sink to this level are so abjectly submissive that the more dominant animals no longer pay any attention to them. Because they no longer interact with other members of the population, they cease to be part of it. Also, their continuing existence is probably the result of an artificial situation created by confinement, as in natural populations such animals would doubtless have been forced to move continually; hence most of them would have become mortality statistics. Such submissive animals have been observed and repeatedly captured in a population of woodchucks.

## CRITICISMS OF THEORY

The criticism has been made, as stated earlier, that results from studies on populations in the laboratory cannot be extrapolated to natural populations because of the excessive densities in the laboratory (11, 60–62). The work cited above (56) showed that mice exposed to crowding for very short periods each day had an increase in adrenal function. In addition, data on density for most natural populations are often misleading, as many species of rodents, especially rats, voles, and mice, often occur in local "colonies" that may be rather crowded even though areas around them may be very sparsely inhabited. Localized groups of rats in natural populations apparently behave like independent populations, with different degrees of crowding, until the numbers and movement increase sufficiently to fill the general area, at which time the colonies lose their identity and become part of a larger population (6). Furthermore, comparable endocrine changes have been observed in natural populations of a number of other species — voles, rats, Japanese deer (sika), woodchucks, and rabbits (30, 53, 63–65). Increased social strife produced by the introduction of aliens into a population of rats will induce movements, increase mortality, and, if the original population was high, cause a striking decline from original densities (18, 66). Conversely, artificial reduction of a population or alteration of its social structure in a way that reduces competition will reduce adrenal weight and incidence of disease accordingly (17, 22, 67). This reduction has been observed in rats, deer, and woodchucks (30, 65, 67).

In some situations no correlation has been shown between adrenocortical function and changes in population, but so far the cases fall into two categories. The first is that where the sample is too small to demonstrate any correlation. For instance, Negus (see 61, Table 4) studied only 98 animals over a two-year period, of all ages and both sexes (61). A second cause of lack of correlation is inaccuracy of population measurement, primarily because currently available census methods are notoriously poor and confidence limits of the estimates have been disregarded (61).

Since changes in adrenal weight occur with reproductive activity, several authors have concluded that adrenal weight cannot be used as an index of adrenocortical function in the study of populations (68, 69). It was implied in these accounts that these changes in adrenal weight with changes in reproductive status were overlooked when conclusions concerning population were drawn from changes in adrenal weight in earlier studies of house mice or other species (62, 68, 70, 71). Our published data show that these factors were taken into consideration in our studies (64, 72). On the other hand, a number of workers may have failed to find a correlation between population status and adrenal weight because changes with sexual function were disregarded. It is well known that adrenal weight increases during pregnancy or with estrogen stimulation in some species, but it is not always remembered that changes in adrenal weight due to population changes can be superimposed on these increases (19, 64, 65, 70). Changes in adrenal weight with change in reproductive status fall into two categories: (i) change in weight when there is immature zonation which later disappears, and (ii) change in weight in fully mature animals which is associated with reproductive activity. Obviously, only changes in adrenal weight or function in animals in the same reproductive condition can be properly compared or correlated with changes in population. Chitty and Clarke (71) have claimed that a marked increase in the size of the adrenal in female voles (*Microtus agrestis*) is restricted to pregnant animals. However, McKeever reported similar increases in nonpregnant, nulliparous, but sexually mature, females of *M. montanus* (69). Our results with *M. pennsylvanicus* are similar, although mature nonpregnant, nulliparous females are scarce, as one would expect. It appears, at least in the North American voles, that the striking increase in the size of the adrenal is associated with maturation and estrogen secretion and is not limited to pregnancy. However, we have long been aware that adrenal size increases during pregnancy in many species and that this must be considered in using adrenal weights as indices of adrenal function. In other cases, a change in adrenal weight is related to seasonal behavioral changes (64), as originally suggested for muskrats (73) and later for *Microtus* (70).

Further criticism of the theory that behavioral-endocrine mechanisms are operative in the control of population growth is based on recent reports of a lack of correlation between adrenal weight and changes in population size, from which it has been concluded that endocrine mechanisms do not affect population growth (61, 68, 69). In another report it was stated that there is no evidence of a "stress mechanism" in a collapse of a lemming population, as no related changes in adrenal weight were found (74). First, it must be noted that failure to demonstrate a correlation, without consideration of pertinent relationships, is not disproof of a correlation. Second, these criticisms have been based on observed adrenal weights in voles or rice rats (*Oryzomys*), primarily in the former without critical evaluation and validation, microscopic or otherwise, of the weight changes. While such conclusions may eventually prove correct in some instances, the inappropriate use, in the studies reported, of adrenal weight as an index of adrenocortical function in these rodents invalidates the conclusions. A basic error in the studies was failure to recognize that many rodents have zones in the adrenal cortex which in many ways resemble the X-zone in house mice, and that these zones are without known function. The use, as indices of function, of the weights of adrenals which include these zones is not appropriate. Delost has published numerous reports on the existence of an "X-zone" in the adrenals of voles (*Clethrionomys*, *Pitymys*, and *Microtus*) which involutes at maturity in males and regenerates during sexual quiescence (see 6). Chitty and Clarke (71) have further explored this problem in *M. agrestis*. On morphological grounds and because we have observed two immature zones in male *M. pennsylvanicus*, we do not entirely agree with Delost that these should be called X-zones, but the basic

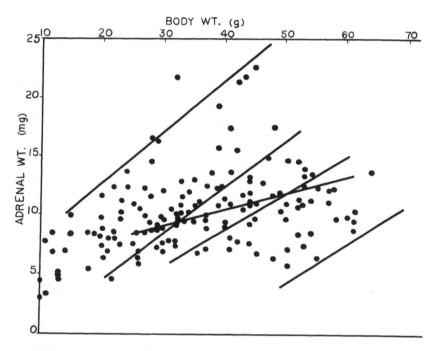

FIGURE 1: A plot of adrenal weights (in milligrams) against body weights (in grams) for male *Microtus pennsylvanicus*. The lines were not fitted but were drawn in approximately as follows. The two parallel lines at upper left enclose points for immature animals whose adrenals showed typical immature zonation. The two parallel lines at lower right enclose points for fully mature animals. The points between (and inevitably to some extent some points enclosed by the parallel lines) are for animals in the process of maturing. The single line corresponds roughly to the mean adrenal weights of maturing animals; however, this is not a fitted line and only suggests the direction of transition. It was found impossible to fit regressions of adrenal weight relative to body weight (or body length) for these data in defining starting or end points of zonal involution.

observation that in immature voles there are zones which later involute, spontaneously or on administration of testosterone, remains valid. Male *M. pennsylvanicus* appears to have two distinct zones which involute at maturity, neither one of which appears to be entirely comparable morphologically to the X-zone of house mice. The male *Pitymys*, *Synaptomys*, and *Clethrionomys* that we have examined, and possibly other voles, have similar zonation, although the probability of differences between species or genera must be kept in mind. These zones persist with inhibition of maturation so that in such voles adrenal weight is relatively much greater in immature than in mature males (Figure 1). The converse is true for the adrenals of female voles (Figures 2–4), which undergo a striking hypertrophy at maturity, probably as an exaggerated effect of estrogens, as described for many species, although this has not been tested as yet (74a). McKeever (69) has demonstrated changes in adrenal weight with age and maturation in *Microtus*, illustrating changes occurring with maturation, but he failed to recognize the zonal changes and probably typical, but enhanced, responses to estrogen, and so arrived at unjustified conclusions. In addition, the picture is confounded by the fact that all gradations between the immature and the fully mature condition of the adrenals occur, as shown in Figures 2 to 5.

Further complicating the picture is the fact that most small mammals born in the fall, and probably even those born at the end of the spring and in the early summer breeding season in a period of relatively high population density, overwinter in the immature condition (17, 19, 75, 76), so that a persistence of immature zonation in males and, in females, the small size of the adrenals unstimulated by estrogen (or whatever factors stimulate the hypertrophy associated with maturation) would be expected (Figures 2 and 3). The basic error in the conclusions of several investigators was the assumption that adrenal weight is always synonymous with cortical function. In addition, nothing is yet known about the steroids secreted by these species or the possible relation of steroid secretory patterns to changes in zonation. Obviously, if adrenal weight is to be used as an index of adrenocortical function in these species, comparisons can be made only between animals in the same state of reproductive function and with the same degree of involution of immature adrenal zones. Thus, for practical purposes, comparisons are

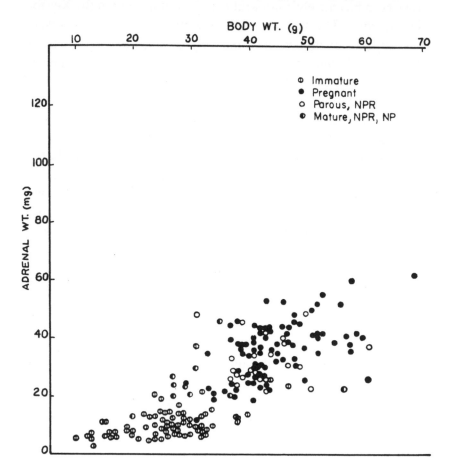

FIGURE 2: A plot of adrenal weights (in milligrams) against body weights (in grams) for female *Microtus pennsylvanicus* captured at all seasons of the year. Reproductive status is indicated as shown (NPR, nonpregnant; NP, nulliparous). It may be seen that adrenal weight increases sharply with sexual maturation whether or not the animal is pregnant or parous, although most were pregnant. As may be seen in Figure 3, most of the immature animals were captured in the late fall and early winter and mainly represent suppressed maturation in young of the preceding breeding season.

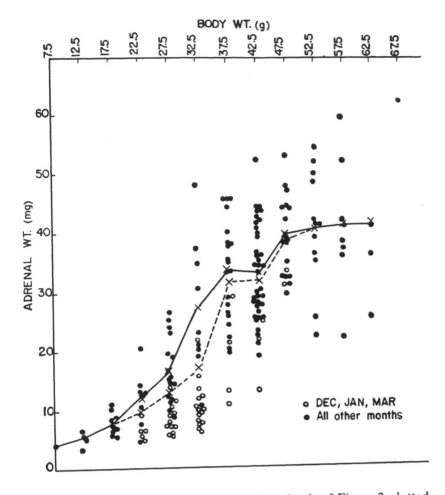

FIGURE 3: Weights of adrenals from the animals of Figure 2 plotted against body weights, with season of capture indicated. A large number of the adrenal weights for immature females are for animals captured between December and March, as indicated in Figure 2. (Solid line): mean adrenal weight for animals captured in any month other than December, January, and March. (Dotted line): mean adrenal weight for all animals. Figures 2 and 3 show that only the weights of adrenals from mature animals can properly be used for comparing changes in weight with changes in population; in the main this means that only the weights of adrenals of animals weighing more than 35 grams can be used, but in the winter one finds a few immature females even in this weight range. For this reason the values for mean adrenal weight that we previously published (18) for female *Microtus montanus* captured in winter are probably too low, although we used only weights of the adrenals of animals weighing 37.5 grams or more in the study, thus largely, but not entirely, avoiding this pitfall.

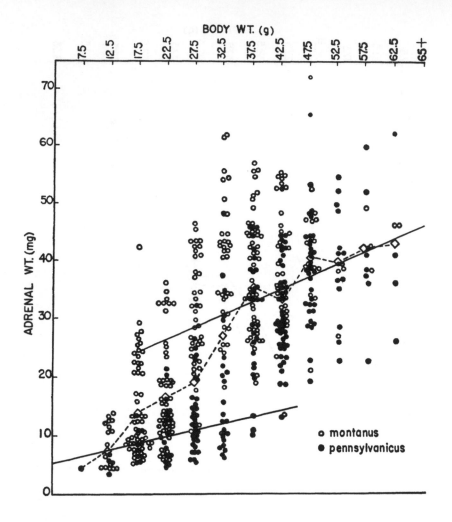

FIGURE 4: Plot of adrenal weights against body weights for female *Microtus pennsylvanicus* and *M. montanus*. Regression curves were fitted to points for fully mature females (upper curve) and to points for immature females (lower curve). Weights of adrenals from maturing females form a continuum between these end stages. This plot again illustrates the problems one encounters in using adrenal weight of voles to assess adrenocortical function in relation to population changes unless one uses only fully mature animals. It appears from this diagram that female *M. montanus* mature somewhat earlier than female *M. pennsylvanicus*, but a number of other factors, including differences in populations, confound the data and make it impossible to draw a definite conclusion.

limited to fully mature, sexually active animals. This means, in our experience, that in *Microtus* (*pennsylvanicus* or *montanus*) one usually is limited to the use of animals weighing 35 to 40 grams or more and having uninhibited reproductive function. The relationship between adrenal weight and reproductive function in terms of the weight of seminal vesicles in *M. pennsylvanicus* is shown in Figure 5. These data illustrate problems one encounters in attempting to make comparisons of adrenal weight for other than fully mature, sexually active males (Figures 1 and 5).

Another cause of failure to demonstrate a correlation between adrenal weight and population density is failure to consider the social rank of individuals in the samples examined. Since high-ranking individuals generally do not have enlarged adrenals,

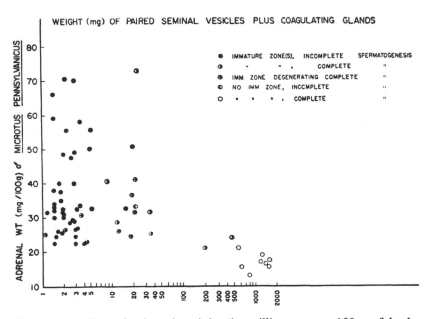

FIGURE 5: Plot of adrenal weight (in milligrams per 100 g of body weight) for male *Microtus pennsylvanicus* against the logarithms of values for combined weights (in milligrams) of seminal vesicles and coagulating glands. The state of sexual maturity is indicated in the key. This plot depicts the difficulties encountered in using adrenal weight to assess adrenocortical function in male microtines because of the presence of immature cortical zones.

comparisons of high-ranking individuals at low and high population densities will reveal no changes in adrenal weight or function. McKeever may have made this error when he divided animals into sexually nonactive and active categories (69). At high population densities maturation of subordinate animals would be delayed, and these animals would be called nonactive. McKeever's Table 2 (69) may simply show that high-ranking animals have similar adrenal weights at low and high population densities, and that low-ranking animals do also. However, comparisons between sexually inactive animals in this case are very probably invalid because of the persistence of immature zonation. Another example of failure to consider social rank is Rudd and Mullen's consideration of only the survivors from groups of pocket gophers (77).

In most instances, failure to find a correlation between adrenal weight and population density is due to inclusion of immature animals in samples in progressively larger numbers with seasonal progression. Seasonal or maturational changes in the adrenals do not invalidate the use of adrenal weights as indicators of adrenocortical function if the weights are used critically, but comparisons must be made between comparable animals at comparable times. For example, it has been possible to show a significant decrease in the size of adrenals of woodchucks with alteration of social structure and diminution of competition during the time of rapid increase in adrenal weight by making the appropriate adjustments (64, 65); however, there is no complicating zonation in this mammal.

So far we have discussed primarily the restriction of increases of populations due to the effects of high densities. In addition, this explanation should provide some understanding of the increased mortality of young that occurs in subsequent generations. One of the striking aspects of both a natural and a confined population is the observation that young animals have a higher mortality than adults coincident with high population and, in natural populations, also after population density has fallen to relatively low levels (7). Chitty (7) explained these losses of young in two ways: losses in the year of peak population he attributed to attacks by adults, and losses in the following spring he attributed to some unknown congenital condition acquired *in utero*. Evidence consistent with this view was presented by Godfrey (78). Body weights that were low as compared with those in the peak year have been reported in these studies (79), and we have observed a similar nonoccurrence of large animals following a peak in population (19).

Chitty (79, 80) invoked genetic selection to explain how later generations might be influenced by conditions existing before they were born. Therefore, he postulates an effect of social behavior different from that proposed by us (17, 19). Thus, he and his colleagues — for example, Krebs (74) and H. Chitty (70) — differ from us on the *kind* of physiological and behavioral changes they postulate, but not on the question of whether behavioral changes play an essential part in the regulation of numbers. We find Chitty's explanation for the events in natural populations difficult to accept because it requires genetic selection acting rapidly for a year or two with a subsequent return to, or close to, the original genetic status. In contrast, we believe there is ample evidence of endocrine mechanisms which have the prolonged effects necessary to account for the increased mortality of young during, and for a considerable time after, episodes of maximum density. We have mentioned some of these effects, but we should also call attention to the life-long effects on reproductive function of single injections of androgens into mice or rats less than 10 days old (see 81); the behavioral effects produced *in utero* reported by Keeley (26); the effects of the injection of corticosteroids or other hormones during pregnancy on later behavior (82); and the effects of corticosteroids on brain development (27). Undernutrition during nursing also has profound and permanent effects on offspring (83), which are consistent with the observed reduced growth at high population levels. Actually neither the endocrine nor the genetic selection explanations have been adequately tested, but there appears to be more evidence in support of the former. However, selection must play a long-range role, if not a short-term one (19). Whatever the mechanisms accounting for the observed increased mortality of young during and following episodes of high density, it seems evident that the altered status, which we believe to be physiological, will increase susceptibility to adverse environmental conditions, and that behavioral factors are of primary importance in the genesis of the altered status. It is clear that, in a general way, we arrive at the same ultimate conclusions as Chitty, but we place more emphasis on decreased productivity than on increased mortality, although one would anticipate various combinations of these two factors to occur in different populations and under different circumstances. One would expect altered reproductive function to be of greater importance in mammals with a high reproductive rate than in those with a much lower reproductive rate, such as woodchucks. Woodchucks exhibit decreased reproductive function with increased social pressure, brought about by an increased failure to mature in their first year, and increased intrauterine mortality (84, 85), but this is less important in regulating their population than movement of young or mortality (17, 84). Of interest in this regard is the finding that young woodchucks become more seriously affected by renal disease at high population densities, and that this probably results in appreciable mortality (17).

These comments lead to consideration of another recent discovery of direct pertinence to the question of the greater effects of high population density on young than on adult mice. First, we repeat that the young in general are subordinate animals and thus, other things being equal, more seriously affected by crowding. However, immature house mice secrete appreciable quantities of 17-hydroxycorticoids, especially hydrocortisone, and, when they are grouped, not only does the total adrenal corticosteroid production increase but the hydrocortisone-corticosterone ratio increases as well (38, 86). With sexual maturation of male mice, the ability to produce hydrocortisone is greatly reduced. Also, if there is delayed maturation accompanying increase in numbers, the secretion of appreciable amounts of hydrocortisone is prolonged. The importance of this finding is that hydrocortisone is a much more potent glucocorticoid than corticosterone, which is the principal compound secreted by adult mice, adult rats, and probably a number of other adult rodents (see 6). Therefore, similar degrees of stimulation of the adrenals of immature and adult mice should result in more profound effects in the immature animals even if there were no difference in social rank. This difference has been observed biologically in the much greater degree of thymic involution and growth suppression produced in immature mice either by ACTH or by grouping than can be produced in adults by similar treatment or by the injection of relatively high amounts of corticosterone (18, 19, 87, 88). Similarly, gonadotrophin secretion is suppressed by much smaller doses of steroids in immature than in mature mice and rats (89), so that inhibition of maturation of the young in experimental and natural populations may be explained on this basis.

In attempting to explain the mechanisms of the progressive inhibition of reproductive function with increasing population density we postulated that increased secretion of adrenal androgens in response to increased secretion of ACTH might be sufficient to inhibit gonadotrophin secretion, especially in immature mice, and thus explain the observed declines in reproductive function (6, 90). Indeed, the injection of adrenal androgens at nonvirilizing physiological concentrations suppresses gonadotrophin secretion and inhibits normal maturation in immature female mice (91). Injection of ACTH in intact immature mice also totally inhibits normal maturation (87). Surprisingly, ACTH has a similar effect in adrenalectomized mice maintained on hydrocortisone (88, 92); thus it appears that ACTH has a direct suppressive effect on reproductive function and therefore on maturation (the site of action is as yet unknown) of immature female mice. Consequently there are at least two distinct mechanisms capable of inhibiting maturation, whose relative importance in the intact animal is unknown. There also remains the distinct possibility that the central nervous system, in response to emotional stimuli, may inhibit gonadotrophin secretion even more directly. In any event, there is ample explanation, including both behavioral and physiological mechanisms, for the differences in the effects of high population levels or increased competition on reproduction, growth, and mortality of the young in contrast to adult animals.

## CONCLUSIONS

The experimental results suggest that there are mechanisms for the regulation of many populations of mammals within the limits imposed by the environment, including food. We subscribe to the view that density-dependent mechanisms have evolved in many forms, and probably in most mammals (11–13, 19, 93). Thus, mammals avoid the hazard of destroying their environment, and thus the hazard of their own extinction. We believe that the evidence, as summarized here, supports the existence of endocrine feedback mechanisms which can regulate and limit population growth in response to increases in over-all "social pressure," and which in turn are a function of increased numbers and aggressive behavior. Neither increased numbers nor increased aggressiveness can operate wholly independently. Furthermore, we believe that environmental factors in most instances probably act through these mechanisms by increasing competition. A good example of this would be the situation described by Errington for muskrats (94). A drought causes the animals to concentrate in areas of remaining water, with the result that competition and social strife are greatly increased. It follows that increased strife, with increased movement, will also increase losses through predation, another way of increasing mortality of subordinate animals (22, 47, 94).

Finally, we might paraphrase Milne's statements (12) regarding density-dependent regulation of population growth as follows: Environmental factors (food, predation, disease, physical factors) may limit population growth, but if they do not, as appears more often than not to be the case in mammals, the physiologic mechanisms outlined above will. And finally, the action of these mechanisms is always proportional to changes that depend on changes in population density, behavior, or both. The fact that a sigmoid growth form requires the operation of such a "density-dependent damping factor" supports this conclusion, whereas external limiting factors, unless they operate through the density-dependent damping mechanism, will characteristically truncate a growth curve. Truncation is seldom seen, but the best example of such a curve for mammals that we have seen is that given by Strecker and Emlen (95).

In summary, we believe that the behavioral-endocrine feedback system is important in the regulation of populations of rodents, lagomorphs, deer, and possibly other mammals. One would expect other factors to occasionally limit population growth, but, when these fail to do so, the feedback mechanism acts as a safety device, preventing utter destruction of the environment and consequent extinction. Because of time-lag effects, this feedback system should not be expected to work perfectly in every situation.

## REFERENCES

1. E. T. Seton, *The Arctic Prairies* (Scribner, New York, 1911).
2. J. R. Dymond, *Trans. Roy. Soc. Can. Sect.* 5, 41, 1 (1947).
3. P. H. Leslie and R. M. Ransom, *J. Animal Ecol.,* 9, 27 (1940).
4. C. H. D. Clarke, *J. Mammal.,* 30, 21 (1949).
5. C. Elton, *Voles, Mice and Lemmings* (Clarendon, Oxford, 1942).
6. J. J. Christian, in *Physiological Mammalogy,* W. V. Mayer and R. G. Van Gelder, eds. (Academic Press, New York, 1963).
7. D. Chitty, *Trans. Roy. Soc. London,* B 236, 505 (1952).
8. D. E. Davis, *Quart. Rev. Biol.,* 28, 373 (1953).
9. P. L. Errington, *Am. Naturalist,* 85, 273 (1951).
10. A. J. Nicholson, *Ann. Rev. Entomol.,* 3, 107 (1958).
11. D. Chitty, *Can. J. Zool.,* 38, 99 (1960).
12. A. Milne, *J. Theoret. Biol.,* 3, 19 (1962).
13. V. C. Wynne-Edwards, *Ibis,* 101, 436 (1959).
14. H. M. Bruce, *J. Reprod. Fertil.,* 1, 96 (1960).
15. C. R. Terman, *Ecol. Bull.,* 44, 123 (1964).
16. J. J. Christian, *J. Mammal.,* 31, 247 (1950).
17. ———, *Military Med.,* 128, 571 (1963).
18. ———, *Proc. Columbia Univ. Symp. Comp. Endocrinol.,* A. Gorbman, ed. (Wiley, New York, 1959), p. 31.
19. ———, *Proc. Natl. Acad. Sci. U.S.,* 47, 428 (1961).
20. I. C. Jones, *The Adrenal Cortex* (Cambridge Univ. Press, Cambridge, 1957).
21. P. V. Rogers and C. P. Richter, *Endocrinology,* 42, 46 (1948).

22. R. Myktowycz, *Australia Commonwealth Sci. Ind. Res. Organ. Wildlife Res.*, 6, 142 (1961).
23. K. Myers and W. E. Poole, *Australian J. Zool.*, 10, 225 (1962).
24. R. L. Helmreich, *Science*, 132, 417 (1960).
25. J. J. Christian and C. D. LeMunyan, *Endocrinology*, 63, 517 (1958).
26. K. Keeley, *Science*, 135, 44 (1962).
27. E. Howard, *Federation Proc.*, 22, 270 (abstr.) (1963).
28. M. R. A. Chance, *Nature*, 177, 228 (1956).
29. L. L. Bernardis and F. R. Skelton, *Proc. Soc. Exptl. Biol. Med.*, 113, 952 (1963).
30. J. J. Christian, V. Flyger, D. E. Davis, *Chesapeake Sci.*, 1, 79 (1960).
31. J. A. Gunn and M. R. Gurd, *J. Physiol. London*, 97, 453 (1940); M. R. A. Chance, *J. Pharmacol. Exptl. Therap.*, 87, 214 (1946); ———, *ibid.*, 89, 289 (1947); E. A. Swinyard, L. D. Clark, J. T. Miyahara, H. H. Wolf, *ibid.*, 132, 97 (1961); G. B. Fink and R. E. Larson, *ibid.*, 137, 361 (1962); R. Ader, A. Kreutner, Jr., H. L. Jacobs, *Psychosomat. Med.*, 25, 60 (1963).
32. E. A. Swinyard, N. Radhakrishnan, L. S. Goodman, *J. Pharmacol. Exptl. Therap.*, 138, 337 (1962); B. Weiss, V. G. Laties, F. L. Blanton, *ibid.*, 132, 366 (1961).
33. J. T. Marsh and A. F. Rasmussen, Jr., *Proc. Soc. Exptl. Biol. Med.*, 104, 180 (1960).
34. D. E. Davis and J. J. Christian, *ibid.*, 94, 728 (1957).
35. J. G. Vandenbergh, *Animal Behavior*, 8, 13 (1960).
36. S. A. Barnett, *Nature*, 175, 126 (1955); K. Eik-Nes, *Record Progr. Hormone Res.*, 15, 380 (1959).
37. F. H. Bronson and B. E. Eleftheriou, *Physiol. Zool.*, 36, 161 (1963).
38. H. H. Varon, J. C. Touchstone, J. J. Christian, *Endocrinology*, in press.
39. W. Eechaute, G. Demeester, E. LaCroix, I. Leusen, *Arch. Intern. Pharmacodyn.*, 136, 161 (1962).
40. A. M. Barrett and M. A. Stockham, *J. Endocrinol.*, 26, 97 (1963).
41. P. G. Pearson, *Bull. Ecol. Soc. Am.*, 43, 134 (abstr.) (1962).
42. J. R. Clarke, *J. Endocrinol.*, 9, 114 (1953).
43. J. Dawson, *Nature*, 178, 1183 (1956).
44. J. P. Rapp and J. J. Christian, *Proc. Soc. Exptl. Biol. Med.*, 114, 26 (1963).
45. R. A. Huseby, F. C. Reed, T. E. Smith, *J. Appl. Physiol.*, 14, 31 (1959); K. A. Khaleque, M. G. Muazzam, R. I. Choadhury, *J. Trop. Med. Hyg.*, 64, 277 (1961); G. G. Slater, R. F. Doctor, E. G. Kollar, paper presented at the 44th meeting of the Endocrine Society (1962).
46. J. J. Christian, *Endocrinology*, 65, 189 (1959).
47. R. M. Lockley, *J. Animal Ecol.*, 30, 385 (1961).
48. C. Kabat, N. E. Collias, R. C. Guettinger, *Wis. Tech. Wildlife Bull.*, No. 7 (1953).
49. L. C. McEwan, C. E. French, N. D. Magruder, R. W. Swift, R. H. Ingram, *Trans. North Am. Wildlife Conf.*, 22, 119 (1957); H. Silver and N. F. Colovos, *Proc. Northeast. Wildlife Conf.*, Portland, Me. (1963).
50. W. Grodzinski, *Proc. Intern. Congr. Zool., 16th* (1963), vol. 1, p. 257.
51. O. Kalela, *Ann. Acad. Sci. Fennicae*, A-IV, No. 34 (1957).
52. A. Gorecki and Z. Gebcaynska, *Acta Theriol.*, 6, 275 (1962).
53. K. Curry-Lindahl, *J. Mammal.*, 43, 171 (1962).
54. J. A. Lloyd and J. J. Christian, *Proc. Intern. Conf. Wildlife Distr., 1st* (1963).
55. B. Welch, *Proc. Intern. Congr. Zool., 16th* (1963), vol. 1, p. 269.
56. F. H. Bronson and B. E. Eleftheriou, *Gen. Comp. Endocrinol.*, 4, 9 (1964).
57. J. J. Christian, *Am. J. Physiol.*, 182, 292 (1955).

58. C. H. Southwick, *Science*, 143, 55 (1964).
59. ———— and V. P. Bland, *Am. J. Physiol.*, 197, 111 (1959).
60. P. Crowcroft and F. P. Rowe, *Proc. Roy. Zool. Soc. London*, 131, 357 (1958).
61. N. C. Negus, E. Gould, R. I. Chipman, *Tulane Studies Zool.*, 8, 95 (1961).
62. R. Tanaka, *Bull. Kochi Women's Univ.*, 10, 7 (1962).
63. B. L. Welch, *Proc. Natl. Deer Distr. Symp. 1st* (Univ. of Georgia Press, Athens, 1962); K. Wodzicki and H. S. Roberts, *New Zealand J. Sci.*, 3, 103 (1960); E. F. Patric, *J. Mammal.*, 43, 200 (1962).
64. J. J. Christian, *Endocrinology*, 71, 431 (1962).
65. J. A. Lloyd, J. J. Christian, D. E. Davis, F. H. Bronson, *Gen. Comp. Endocrinol.*, 4, 271 (1964).
66. D. E. Davis, *Trans. North. Am. Wildlife Conf., 14th* (1949), p. 225.
67. J. J. Christian and D. E. Davis, *Trans. North. Am. Wildlife Conf., 20th* (1955), p. 177.
68. D. A. Mullen, *J. Mammal.*, 41, 129 (1960).
69. S. McKeever, *Anat. Record*, 135, 1 (1959).
70. H. Chitty, *J. Endocrinol.*, 22, 387 (1961).
71. ———— and J. R. Clarke, *Can. J. Zool.*, 41, 1025 (1963).
72. J. J. Christian, *Ecology*, 37, 258 (1956).
73. J. R. Beer and R. K. Meyer, *J. Mammal.*, 32, 173 (1951).
74. C. J. Krebs, *Science*, 140, 674 (1963).
74a. Results from a recent study of the relationships between sexual maturity, the adrenal glands, and populations density in female *M. pennsylvanicus* from a natural population suggest that *female* voles of this species have no X-zone as it is defined for house mice (20). Apparently it is a hypertrophic reticularis and inner fasciculata [resembling the adrenals of woodchucks in this respect (54, 64)] which have been labeled an X-zone. There is no involution in pregnancy. The cells contain lipids, and the hyperplasia occurs with maturation, as pointed out by Chitty and Clarke (71). Adrenal weight relative to body weight is a discontinuous function in these animals as a result of the sudden increase at maturation. Therefore, regressions of adrenal weight relative to body weight or body length are invalid if the data come from both immature and mature females. When these facts are taken into account it is clear that there is no change in adrenal weight relative to body weight with reproductive status in mature females, and that there is a remarkable parallelism between mean adrenal weight of mature females and population size.
75. A. van Wijngaarden, *Verslag Landbouwk. Onderzoek No. 66.22* (1960), pp. 1–68.
76. D. A. Spencer "The Oregon Meadow Mouse Irruption of 1957–58," *Fed. Coop. Expt. Serv., Corvallis, Publ.* (1959), p. 15; K. A. Adamczewska, *Acta Theriol.*, 5, 1 (1961); W. Sheppe, *J. Mammal.*, 44, 180 (1963); D. R. Breakey, *ibid.*, p. 153.
77. R. L. Rudd and D. A. Mullen, *J. Mammal.*, 44, 451 (1963).
78. G. R. Godfrey, *ibid.*, 36, 209 (1955).
79. D. Chitty, *Cold Spring Harbor Symp. Quant. Biol.*, 22, 277 (1958).
80. H. Chitty and D. Chitty, *Symp. Theriol. Prague* (Czechoslovak Academy of Science, Prague, 1962), p. 77.
81. R. A. Gorski and C. A. Barraclough, *Endocrinology*, 73, 210 (1963).
82. M. W. Lieberman, *Science*, 141, 824 (1963).
83. E. M. Widdowson and G. C. Kennedy, *Proc. Roy. Soc. London*, B 156, 96 (1962); E. M. Widdowson and R. A. McCance, *ibid.*, B 158, 329 (1963).

84. D. E. Davis, *J. Wildlife Management*, 26, 144 (1962).
85. R. L. Snyder, *Ecology*, 43, 506 (1962).
86. H. H. Varon, J. C. Touchstone, J. J. Christian, *Federation Proc.*, 22, 164 (abstr.) (1963).
87. J. J. Christian, *Endocrinology*, 74, 669 (1964).
88. ———, *ibid.*, 75, 653 (1964).
89. W. W. Byrnes and R. K. Meyer, *ibid.*, 48, 133 (1951); W. W. Byrnes and E. G. Shipley, *Proc. Soc. Exptl. Biol. Med.*, 74, 308 (1950); D. Ramirez and S. M. McCann, *Endocrinology*, 72, 452 (1963).
90. J. J. Christian, *Proc. Soc. Exptl. Biol. Med.*, 104, 330 (1960).
91. H. H. Varon and J. J. Christian, *Endocrinology*, 72, 210 (1963); G. E. Duckett, H. H. Varon, J. J. Christian, *ibid.*, p. 403.
92. J. J. Christian, *Federation Proc.*, 22, 507 (abstr.) (1963).
93. F. A. Pitelka, *Cold Spring Harbor Symp. Quant. Biol.*, 22, 237 (1958).
94. P. L. Errington, *Agr. Expt. Sta. Iowa State Coll., Ames, Res. Bull.*, 320 (1943).
95. R. L. Strecker and J. T. Emlen, *Ecology*, 35, 249 (1953).

# 7. Self-Regulating Systems in Populations of Animals

## V. C. WYNNE-EDWARDS

I am going to try to explain a hypothesis which could provide a bridge between two biological realms.[1] On one side is that part of the "Balance of Nature" concerned with regulating the numbers of animals, and on the other is the broad field of social behavior. The hypothesis may, I believe, throw a bright and perhaps important sidelight on human behavior and population problems. I must emphasize, however, that it is still a hypothesis. It appears to be generally consistent with the facts, and it provides entirely new insight into many aspects of animal behavior that have hitherto been unexplainable; but because it involves long-term evolutionary processes it cannot be put to an immediate and comprehensive test by short-term experiments.

Human populations are of course increasing at compound interest practically all over the world. At the over-all 2 per cent

From *Science*, 147 (March 26, 1965), 1543–1548. Copyright 1965 by the American Association for the Advancement of Science. The article is based on a lecture presented December 26, 1964, at the Montreal meeting of the AAAS.

annual rate of the last decade, they can be expected to double with each generation. In the perspective of evolutionary time such a situation must be extremely short-lived, and I am sure we are going to grow more and more anxious about the future of man until we are able to satisfy ourselves that the human population explosion is controllable, and can be contained.

Populations of animals, especially when they are living under primeval undisturbed conditions, characteristically show an altogether different state of affairs; and this was equally true of man in the former cultural periods of the stone age. These natural populations tend to preserve a continuing state of balance, usually fluctuating to some extent but essentially stable and regulated. The nature of the regulatory process has been the main focus of study and speculation by animal ecologists during the whole of my working life, and in fact considerably longer.

Charles Darwin[2] was the first to point out that though all animals have the capacity to increase their numbers, in fact they do not continuously do so. The "checks to increase" appeared to him to be of four kinds — namely, the amount of food available, which must give the extreme limit to which any species can increase; the effects of predation by other animals; the effects of physical factors such as climate; and finally, the inroads of disease. "In looking at Nature," he tells us in the *Origin of Species*, "it is most necessary . . . never to forget that every single organic being may be said to be striving to the utmost to increase in numbers." This intuitive assumption of a universal resurgent pressure from within held down by hostile forces from without has dominated the thinking of biologists on matters of population regulation, and on the nature of the struggle for existence, right down to the present day.

Setting all preconceptions aside, however, and returning to a detached assessment of the facts revealed by modern observation and experiment, it becomes almost immediately evident that a very large part of the regulation of numbers depends not on Darwin's hostile forces but on the initiative taken by the animals themselves; that is to say, to an important extent it is an intrinsic phenomenon.

Forty years ago Jespersen[3] showed, for example, that there is a close numerical agreement between the standing crop of planktonic organisms at the surface of the North Atlantic Ocean and the distribution density of the various deep-sea birds that depend on these organisms for food. Over the whole of this vast area the oceanic birds are dispersed in almost constant proportion to the local biomass of plankton, although the biomass itself varies from region to region by a factor of about 100; the actual crude correlation coefficient is 85 per cent. This pro rata dispersion of the birds must in fact depend solely on their own intrinsic efforts and behavior. Even though the dispersion directly reflects the availability of food, the movements of the birds over the ocean are essentially voluntary and not imposed against their will by hostile or other outside forces.

Turning to the results of repeatable experiments with laboratory animals, it is a generally established principle that a population started up, perhaps from one parental pair, in some confined universe such as an aquarium or a cage, can be expected to grow to a predictable size, and thereafter to maintain itself at that ceiling for months or years as long as the experimenter keeps the conditions unchanged. This can readily be demonstrated with most common laboratory animals, including the insects *Drosophila* and *Tribolium*, the water-flea *Daphnia*, the guppy *Lebistes*, and also mice and rats. The ceiling population density stays constant in these experiments in the complete absence of predators or disease and equally without recourse to regulation by starvation, simply by the matching of recruitment and loss. For example, a set of particularly illuminating experiments by Silliman and Gutsell,[4] lasting over three years, showed that when stable populations of guppies, kept in tanks, were cropped by removal of a proportion of the fish at regular intervals, the remainder responded by producing more young that survived, with the consequence that the losses were compensated. In the controls, on the other hand, where the stocks were left untouched, the guppies went on breeding all the time, but by cannibalism they consistently removed at birth the whole of the surplus produced. The regulating methods are different in different species; under appropriate circumstances in mice, to take another example, ovulation and reproduction can decline and even cease, as long as the ceiling density is maintained.

Here again, therefore, we are confronted by intrinsic mechanisms, in which none of Darwin's checks play any part, competent in themselves to regulate the population size within a given habitat.

The same principle shows up just as clearly in the familiar concept that a habitat has a certain carrying capacity, and that it is no good turning out more partridges or planting more trout than the available habitat can hold.

Population growth is essentially a density-dependent process; this means that it tends to proceed fastest when population densities are far below the ceiling level, to fall to zero as this level is approached, and to become negative, leading to an actual drop in numbers, if ever the ceiling is exceeded. The current hypothesis is that the adjustment of numbers in animals is a homeostatic process — that there is, in fact, an automatic self-righting balance between population density and resources.

I must turn briefly aside here to remind you that there are some environments which are so unstable or transitory that there is not time enough for colonizing animals to reach a ceiling density, and invoke their regulatory machinery, before the habitat becomes untenable again or is destroyed. Populations in these conditions are always in the pioneering stage, increasing freely just as long as conditions allow. Instability of this kind tends to appear around the fringes of the geographical range of all free-living organisms, and especially in desert and polar regions. It is also very common in agricultural land, because of the incessant disturbance of ploughing, seeding, spraying, harvesting, and rotating of crops. In these conditions the ecologist will often look in vain for evidences of homeostasis, among the violently fluctuating and completely uncontrollable populations typical of the animal pests of farms and plantations. Homeostasis can hardly be expected to cope unerringly with the ecological turmoil of cultivated land.

I return later to the actual machinery of homeostasis. For the present it can be accepted that more or less effective methods of regulating their own numbers have been evolved by most types of animals. If this is so, it seems logical to ask as the next question: What is it that decides the ceiling level?

## FOOD SUPPLY AS A LIMITING FACTOR

Darwin was undoubtedly right in concluding that food is the factor that normally puts an extreme limit on population density, and the dispersion of oceanic birds over the North Atlantic, which so closely reflects the dispersion of their food supply, is certain to prove a typical and representative case. Just the same, the link between food productivity and population density is very far from being self-evident. The relationship between them does not typically involve any signs of undernourishment; and starvation, when we observe it, tends to be a sporadic or accidental cause of mortality rather than a regular one.

Extremely important light is shed on this relationship between population density and food by our human experience of exploiting resources of the same kind. Fish, fur-bearing animals, and game are all notoriously subject to overexploitation at the hands of man, and present-day management of these renewable natural resources is based on the knowledge that there is a limit to the intensity of cropping that each stock can withstand. If we exceed this critical level, the stock will decline and the future annual crops will diminish. Exactly parallel principles apply to the exploitation of natural prairie pastures by domestic livestock: if overgrazing is permitted, fertility and future yields just as fatally decline.

In all these situations there is a tendency to overstep the safety margin while exploitation of the resource is still economically profitable. We have seen since World War II, for example, the decimation of stocks of the blue and the humpback whale in the southern oceans, under the impetus of an intense profit motive, which persisted long after it had become apparent to everyone in the industry that the cropping rate was unsupportably high. The only way to protect these economically valuable recurrent resources from destruction is to impose, by agreement or law, a manmade code of rules, defining closed seasons, catch limits, permitted types of gear, and so on, which restrict the exploitation rate sufficiently to prevent the catch from exceeding the critical level.

In its essentials, this is the same crucial situation that faces populations of animals in exploiting their resources of food. Indeed, without going any further one could predict that if the food supplies of animals were openly exposed to an unruly scramble, there could be no safeguard against their over-exploitation either.

## CONVENTIONAL BEHAVIOR IN RELATION TO FOOD

When I first saw the force of this deduction ten years ago, I felt that the scales had fallen from my eyes. At once the vast edifice of conventional behavior among animals in relation to food began to take on a new meaning. A whole series of unconnected natural phenomena seemed to click smoothly into place.

First among these are the territorial systems of various birds (paralleled in many other organisms), where the claim to an individual piece of ground can evoke competition of an intensity unequaled on any other occasion in the life of the species concerned. It results, in the simplest cases, in a parceling out of the habitat into a mosaic of breeding and feeding lots. A territory has to be of a certain size, and individuals that are unsuccessful in obtaining one are often excluded completely from the habitat, and always prevented from breeding in it. Here is a system that might have been evolved for the exact purpose of imposing a ceiling density on the habitat, and for efficiently disposing of any surplus individuals that fail to establish themselves. Provided the territory size is adequate, it is obvious that the rate of exploitation of the food resources the habitat contains will automatically be prevented from exceeding the critical threshold.

There are other behavioral devices that appear, in the light of the food-resource hypothesis we are examining, equally purposive in leading to the same result — namely, that of limiting the permitted quota of participants in an artificial kind of way, and of off-loading all that are for the time being surplus to the carrying capacity of the ground. Many birds nest in colonies — especially, for example, the oceanic and aerial birds which cannot, in the nature of things, divide up the element in which they feed into static individual territories. In the colony the pairs compete just as long and keenly for one of the acceptable nest sites, which are in some instances closely packed together. By powerful tradition some of these species return year after year to old-established resorts, where the perimeter of the colony is closely drawn like an imaginary fence around the occupied sites. Once again there is not always room to accommodate all the contestants, and unsuccessful ones have to be relegated to a nonbreeding surplus or reserve, inhibited from sexual maturation because they have failed to obtain a site within the traditional zone and all other sites are taboo.

A third situation, exemplifying another, parallel device, is the pecking order or social hierarchy so typical of the higher animals that live in companies in which the individual members become mutually known. Animal behaviorists have studied the hierarchy in its various manifestations for more than forty years, most commonly in relation to food. In general, the individuals of higher rank have a prior right to help themselves, and, in situations where there is not enough to go round, the ones at the bottom of the scale must stand aside and do without. In times of food shortage — for example, with big game animals — the result is that the dominant individuals come through in good shape while the subordinates actually die of starvation. The hierarchy therefore produces the same kind of result as a territorial system in that it admits a limited quota of individuals to share the food resources and excludes the extras. Like the other devices I have described, it can operate in exactly the same way with respect to reproduction. In fact, not only can the hierarchical system exclude individuals from breeding, it can equally inhibit their sexual development.

It must be quite clear already that the kind of competition we are considering, involving as it does the right to take food and the right to breed, is a matter of the highest importance to the individuals that engage in it. At its keenest level it becomes a matter of life and death. Yet, as is well known, the actual contest between individuals for real property or personal status is almost always strictly conventionalized. Fighting and bloodshed are superseded by mere threats of violence, and threats in their turn are sublimated into displays of magnificence and virtuosity. This is the world of bluff and status symbols. What takes place, in other words, is a contest for conventional prizes conducted under conventional rules. But the contest itself is no fantasy, for the losers can forfeit the chance of posterity and the right to survive.

It is at this point that the hypothesis provides its most unexpected and striking insight, by showing that the conventionalization of rivalry and the foundation of society are one and the same thing. Hitherto it has never been possible to give a scientific definition of the terms *social* and *society*, still less a functional explanation. The emphasis has always been on the rather vague element of companionship and brotherhood. Animals have in the main been regarded as social whenever they were gregarious. Now we can view the social phenomenon in a new light. According to the hypothesis the society is no more and no less than the organization necessary for the staging of conventional competition. At once it assumes a crisp definition: a society is an organization of individuals that is capable of providing conventional competition among its members.

Such a novel interpretation of something that involves us all so intimately is almost certain to be viewed at first sight a bit skeptically; but in fact one needs no prompting in our competitive world to see that human society is impregnated with rivalry. The sentiments of brotherhood are warm and reassuring, and in identifying society primarily with these we appear to have been unconsciously shutting our eyes to the inseparable rough-and-tumble of status seeking and social discrimination that are never very far to seek below the surface, bringing enviable rewards to the successful and pitiful distress to those who lose. If this interpretation is right, conventional competition is an inseparable part of the substance of human society, at the parochial, national, and international level. To direct it into sophisticated and acceptable channels is one of the great motives of civilized behavior; but it would be idle to imagine that we could eliminate it.

A corollary of the hypothesis that deserves mention is the extension of sociality that it implies, to animals of almost every kind whether they associate in flocks or seek instead a more solitary way of life. There is no particular difficulty of course in seeing, for example, cats and dogs as social mammals individually recognizing the local and personal rights of acquaintances and strangers and inspired by obviously conventional codes of rivalry when they meet. In a different setting, the territory-holding birds that join in the chorus of the spring dawn are acting together in social concert, expressing their mutual rivalry by a conventional display of exalted sophistication and beauty. Even at the other extreme, when animals flock into compact and obviously social herds and schools, each individual can sometimes be seen to maintain a strict individual distance from its companions.

## Social Organization and Feedback

We can conveniently return now to the subject of homeostasis, in order to see how it works in population control. Homeostatic systems come within the general purview of cybernetics; in fact, they have long been recognized in the physiology of living organisms. A simple model can be found in any thermostatic system, in which there must of course be units capable of supplying or withdrawing heat whenever the system departs from its standard temperature and readjustment is necessary. But one also needs an indicator device to detect how far the system has deviated and in which direction. It is the feedback of this information that activates the heating or cooling units.

Feedback is an indispensable element of homeostatic systems. There seems no reason to doubt that, in the control of population density, it can be effectively provided simply by the intensity of conventional competition. Social rivalry is inherently density-dependent: the more competitors there are seeking a limited number of rewards, the keener will be the contest. The impact of stress on the individuals concerned, arising from conventional competition and acting through the pituitary-adrenal system, is already fully established, and it can profoundly influence their responses, both physiological and behavioral.

One could predict on theoretical grounds that feedback would be specially important whenever a major change in population density has to take place, upsetting the existing balance between demand and resources. This must occur particularly in the breeding season and at times of seasonal migrations. Keeping this in mind, we can obtain what we need in the way of background information by examining the relatively long-lived vertebrates, including most kinds of birds and mammals, whose individual members live long enough to constitute a standing population all the year round. The hypothesis of course implies that reproduction, as one of the principal parameters of population, will be subject to control — adjusted in magnitude, in fact, to meet whatever addition is currently required to build up the population and make good the losses of the preceding year. *Recruitment* is a term best used only to mean intake of new breeding adults into the population, and in that sense, of course, the raw birth rate may not be the sole and immediate factor that determines it. The newborn young have got to survive adolescence before they can become recruits to the breeding stock; and even after they attain puberty, social pressures may exclude them from reproducing until they attain a sufficiently high rank in the hierarchy. Indeed, there is evidence in a few species that, under sufficient stress, adults which have bred in previous years can be forced to stand aside.

There are, in fact, two largely distinct methods of regulating reproductive output, both of which have been widely adopted in the animal kingdom. One is to limit the number of adults that are permitted to breed, and this is of course a conspicuous result of adopting a territorial system, or any other system in which the number of permissible breeding sites is restricted. The other is to influence the number of young that each breeding pair is conditioned to produce. The two methods can easily be combined.

What we are dealing with here is a part of the machinery for adjusting population density. What we are trying to get at, however, is the social feedback mechanism behind it, by which the appropriate responses are elicited from potential breeders.

Birds generally provide us with the best examples, because their size, abundance, and diurnal habits render them the most observable and familiar of the higher animals. It is particularly easy to see in birds that social competition is keenest just before and during the breeding season, regardless of the type of breeding dispersion any given species happens to adopt. Individuals may compete for and defend territories or nest sites, or in rarer cases they may engage in tournaments in an arena or on a strutting ground; and they may join in a vocal chorus especially concentrated about the conventional hours of dawn and dusk, make mass visits to colony sites, join in massed flights, and share in other forms of communal displays. Some of these activities are more obviously competitive than others, but all appear to be alike in their capacity to reveal to each individual the concentration or density level of the population within its own immediate area.

## COMMUNAL MALE DISPLAYS

Some of these activities, like territorial defense, singing, and the arena displays, tend to be the exclusive concern of the males. It has never been possible hitherto to give a satisfactory functional explanation of the kind of communal male displays typified by the arena dances of some of the South American hummingbirds and manakins, and by the dawn strutting of prairie chickens and sharp-tailed grouse. The sites they use are generally traditional, each serving as a communal center and drawing the competitors from a more or less wide surrounding terrain. On many days during the long season of activity the same assembly of males may engage in vigorous interplay and mutual hostility, holding tense dramatic postures for an hour or more at a stretch without a moment's relaxation, although there is no female anywhere in sight at the time. The local females do of course come at least once to be fertilized; but the performance makes such demands on the time and energy of the males that it seems perfectly reasonable to assume that this is the reason why they play no part in nesting and raising a family. The duty they perform is presumably important, but it is simply not credible to attribute it primarily to courting the females. To anyone looking for a population feedback device, on the other hand, interpretation would present no difficulty: he would presume that the males are being conditioned or stressed by their ritual exertions. In some of the arena species some of the males are known to be totally excluded from sexual intercourse; but it would seem that the feedback mechanism could produce its full effect only if it succeeded in limiting the number of females fertilized to an appropriate quota, after which the males refused service to any still remaining unfertilized. I hope research may at a not-too-distant date show us whether or not such refusal really takes place.

The conclusion that much of the social display associated with the breeding season consists of males competing with males makes necessary a reappraisal of Darwinian sexual selection. Whether the special organs developed for display are confined to the males, as in the examples we have just considered, or are found in both sexes, as for instance in most of the colony-nesting birds, there is a strong indication that they are first and foremost status symbols, used in conventional competition, and that the selective process by which they have been evolved is social rather than sexual. This would account for the hitherto puzzling fact that although in the mature bullfrog and cicada the loud sound is produced by the males, in both cases it is the males that are provided with extra-large eardrums. There does not seem much room for doubt about who is displaying to whom.

Communal displays are familiar also in the context of bird migration, especially in the massing and maneuvering of flocks before the exodus begins. A comparable buildup of social excitement precedes the migratory flight of locusts. Indeed, what I have elsewhere defined as *epideictic* phenomena — displays, or special occasions, which allow all the individuals taking part to sense or become conditioned by population pressure — appear to be very common and widespread in the animal kingdom. They occur especially at the times predicted, when feedback is required in anticipation of a change in population density. The singing of birds, the trilling of katydids, crickets, and frogs, the underwater sounds of fish, and the flashing of fireflies all appear to perform this epideictic function. In cases where, as we have just seen, epideictic behavior is confined in the breeding season to the male sex, the presumption is that the whole process of controlling the breeding density and the reproductive quota is relegated to the males. Outside the breeding season, when the individuals are no longer united in pairs and are all effectively neuter in sex, all participate alike in epideictic displays — in flighting at sundown, like ducks; in demonstrating at huge communal roosts at dusk, like starlings, grackles, and crows; or in forming premigratory swarms, like swallows. The assumption which the hypothesis suggests, that the largest sector of all social behavior must have this fundamentally epideictic or feedback function, gives a key to understanding a vast agglomeration of observed animal behavior that has hitherto been dubiously interpreted or has seemed altogether meaningless.

## Maintaining Population Balance

Having outlined the way in which social organization appears to serve in supplying feedback, I propose to look again at the machinery for making adjustments to the population balance. In territorial birds, variations in the average size of territories from place to place and year to year can be shown to alter the breeding density and probably also the proportion of adults actually participating in reproduction. In various mammals the proportion of the females made pregnant, the number and size of litters, the survival of the young and the age at which they mature may all be influenced by social stress. Wherever parental care of the young has been evolved in the animal kingdom, the possibility exists that maternal behavior and solicitude can be affected in the same way; and the commonly observed variations in survival rates of the newborn could, in that case, have a substantial functional component and play a significant part in regulating the reproductive output. This would, among other things, explain away the enigma of cannibalism of the young, which we noticed earlier in the guppies and which occurs sporadically all through the higher animals. Infanticide played a conspicuous part in reducing the effective birth rate of many of the primitive human peoples that survived into modern times. Not infrequently it took the form of abandoning the child for what appeared to be commendable reasons, without involving an act of violence.

Reproduction is of course only one of the parameters involved in keeping the balance between income and loss in populations. The homeostatic machinery can go to work on the other side of the balance also, by influencing survival. Already, in considering the recruitment of adults, we have taken note of the way this can be affected by juvenile mortality, some of which is intrinsic in origin and capable of being promoted by social pressures. Conventional competition often leads to the exclusion of surplus individuals from any further right to share the resources of the habitat, and this in turn compels them to emigrate. Research conducted at Aberdeen in recent years has shown how important a factor forced expulsion is in regulating the numbers of the Scottish red grouse. Every breeding season so far has produced a population surplus, and it is the aggressive behavior of the dominant males which succeeds in driving the supernumeraries away. In this case the outcasts do not go far; they get picked up by predators or they mope and die because they are cut off from their proper food. Deaths from predation and disease can in fact be substantially "assisted" under social stress.

On the income side, therefore, both reproductive input and the acquisition of recruits by immigration appear to be subject to social regulation; and on the loss side, emigration and what can be described as socially induced mortality can be similarly affected. Once more it appears that it is only the inroads of Darwin's "checks to increase," the agents once held to be totally responsible for population regulation, which are in fact uncontrollable and have to be balanced out by manipulation of the other four components.

Attention must be drawn to the intimate way in which physiology and behavior are entwined in providing the regulatory machinery. It seems certain that the feedback of social stimulation acts on the individual through his endocrine system, and in the case of the vertebrates, as I have said, this particularly involves the pituitary and adrenal cortex or its equivalent. Sometimes the individual's response is primarily a physiological one — for example, the inhibition of spermatogenesis or the acceleration of growth; sometimes it is purely behavioral, as in the urge to return to the breeding site, the development of aggressiveness, or the demand for territory of a given size. But often there is a combination of the two — that is to say, a psychosomatic response, as when, for instance, the assumption of breeding colors is coupled with the urge to display.

## SOURCES OF CONTROVERSY

There is no need for me to emphasize that the hypothesis is controversial. But almost all of it is based on well-established fact, so that the controversy can relate solely to matters of interpretation. Examples have been given here which show the ability of the hypothesis to offer new and satisfying interpretations of matters of fact where none could be suggested before. Some of these matters are of wide importance, like the basic function of social behavior; some are matters of everyday experience, like why birds sing at dawn. Very seldom indeed does the hypothesis contradict well-founded accepted principles. What, then, are the sources of controversy?

These are really three in number, all of them important. The first is that the concept is very wide-ranging and comprehensive; this means that it cannot be simply proved or disproved by performing a decisive experiment. There are of course dubious points where critical tests can be made, and research is proceeding, at Aberdeen among many other places, toward this end. Relevant results are constantly emerging, and at many points the hypothesis has been solidified and strengthened since it was first formulated. On the other hand, there has been no cause yet to retract anything.

The second source of controversy is that the hypothesis invokes a type of natural selection which is unfamiliar to zoologists generally. Social grouping is essentially a localizing phenomenon, and an animal species is normally made up of countless local populations all perpetuating themselves on their native soil, exactly as happens in underdeveloped and primitive communities of man. Social customs and adaptations vary from one local group to another, and the hypothesis requires that natural selection should take place between these groups, promoting those with more effective social organizations while the less effective ones go under. It is necessary, in other words, to postulate that social organizations are capable of progressive evolution and perfection as entities in their own right. The detailed arguments [5] are too complex to be presented here, but I can point out that intergroup selection is far from being a new concept: It has been widely accepted for more than twenty years by geneticists. It is almost impossible to demonstrate it experimentally because we have to deal with something closely corresponding to the rise and fall of nations in history, rather than with success or failure of single genes over a few generations; it is therefore the time scale that prevents direct experiment. Even the comparatively rapid process of natural selection acting among individuals has been notoriously difficult to demonstrate in nature.

The third objection is, I think, by far the most interesting. It is simply that the hypothesis does not apply to ourselves. No built-in mechanisms appear to curb our own population growth, or adjust our numbers to our resources. If they did so, everything I have said would be evident to every educated child, and I should not be surveying it here. How is this paradox to be explained?

The answer, it seems clear, is that these mechanisms did exist in primitive man and have been lost, almost within historic times. Man in the paleolithic stage, living as a hunter and gatherer, remained in balance with his natural resources just as other animals do under natural conditions. Generation after generation, his numbers underwent little or no change. Population increase was prevented not by physiological control mechanisms of the kind found in many other mammals but only by behavioral ones, taking the form of traditional customs and taboos. All the stone age tribes that survived into modern times diminished their effective birth rate by at least one of three ritual practices — infanticide, abortion, and abstention from intercourse. In a few cases, fertility was apparently impaired by surgery during the initiation ceremonies. In many cases, marriage was long deferred. Mortality of those of more advanced age was often raised through cannibalism, tribal fighting, and human sacrifice.

Gradually, with the spread of the agricultural revolution, which tended to concentrate the population at high densities on fertile soils and led by degrees to the rise of the town, the craftsman, and the merchant, the old customs and taboos must have been forsaken. The means of population control would have been inherited originally from man's subhuman ancestors, and among stone age peoples their real function was probably not even dimly discerned except perhaps by a few individuals of exceptional brilliance and insight. The continually expanding horizons and skills of modern man rendered intrinsic limitation of numbers unnecessary, and for 5,000 or 10,000 years the advanced peoples of the Western world and Asia have increased without appearing to harm the world about them or endanger its productivity. But the underlying principles are the same as they have always been. It becomes obvious at last that we are getting very near the global carrying capacity of our habitat, and that we ought swiftly to impose some new, effective, homeostatic regime before we overwhelm it, and the ax of group selection falls.

REFERENCES

1. V. C. Wynne-Edwards, *Animal Dispersion in Relation to Social Behaviour* (Hafner, New York, 1962).
2. C. Darwin, *The Origin of Species* (Murray, London, 1859) (quoted from 6th edition, 1872).
3. P. Jespersen, "The frequency of birds over the high Atlantic Ocean," *Nature*, 114, 281 (1924).
4. R. P. Silliman and J. S. Gutsell, "Experimental exploitation of fish populations," *U.S. Fish Wildlife Serv. Fishery Bull.*, 58, 214 (1958).
5. V. C. Wynne-Edwards, "Intergroup selection in the evolution of social systems," *Nature*, 200, 623 (1963).

# 8: On Group Selection and Wynne-Edwards' Hypothesis

## JOHN A. WIENS

The means by which the sizes of animal populations are regulated, or whether in fact any real regulation of numbers occurs at all, have concerned ecologists and evolutionists for many years, and have engendered much controversy. Many have accepted Lack's view (1954) that the number of individuals in a population is directly limited by (and balanced with) the food supply of an environment, some animals simply starving when the numbers have exceeded the food resources. Others have taken exception to this view, and have proposed rather that the essential control consists in preventing food shortages by limiting population size at a level commensurate with the resources of the environment. This view has been championed especially by the Finnish workers Kalela (1957) and Koskimies (1955) and has been most extensively elaborated by Wynne-Edwards in his stimulating book *Animal Dispersion in Relation*

From *American Scientist,* 54 (1966), 273–287.

*to Social Behavior* (1962), in which he has developed a hypothesis which incorporates a diverse array of social behaviors within the single functional categorization of population regulation. Briefly stated, his hypothesis proposes that, while food is generally the environmental resource which ultimately limits animal populations, free contest among individuals for food must in the long run end in overexploitation of the resource and damage the habitat or lead to mass starvation. This situation is usually avoided by the substitution of "conventional rewards," such as territories, living space, social rank, etc., in place of the actual food itself. Competition for these conventional rewards operates homeostatically, adjusting the population density at the "optimum" level in relation to fluctuating levels of food resources and thereby preventing increases to densities that would cause overexploitation and the depletion of future yields. Many social displays are "epideictic," serving to indicate the density of individuals present in a locality, and thus provide the feedback to adjust or restore the appropriate balance between population density and consumable resources. Much of conventional competition involves subordination of the welfare of individuals to that of the social group to which they belong. Such mechanisms could only have arisen through a process of "group selection," in which populations with the appropriate behavioral mechanisms would regulate their density and persist, populations lacking such behavior overexploiting their resources and perishing.

The ideas Wynne-Edwards has expressed in this hypothesis have met with varied responses, some favorable, and many unfavorable. Wynne-Edwards considers the process of group selection to be fundamental to his hypothesis, an essential part of its framework, but it is this aspect which has evoked perhaps the most criticism and caused the most controversy. Elton (1963) proposes that "the enormous weakness of this enormous book" is the lack of verification of group selection, and Brown (1964) holds that Wynne-Edwards' insistence on group selection invalidates his argument relating the evolution of territoriality to population regulation. In his well-considered discussion,

Crook (1965) has raised a number of objections to the hypothesis, particularly its dependence on group selection, and proposes that social behavior as a mechanism of regulating population size should be examined through an approach based upon orthodox evolutionary principles rather than group selection. Others (e.g., Amadon, 1964; Orians and Willson, 1964; Maynard Smith, 1964; King, 1965; Selander, 1965; Lack, 1965; Tinbergen, 1965) have pointed out that group selection is either untenable or unnecessary to explain the evolution of various "regulatory" mechanisms. A rather different view is expressed by Braestrup (1963a, 1963b), who has serious objections to ascribing any regulatory function to social behavior, but is in perfect accord with Wynne-Edwards on the necessity of group selection for the evolution of such behavior, and in fact feels that much of this behavior may have evolved to facilitate group selection, since it often maintains the localization of a group; this, as we shall see, is a necessary ingredient of group selection.

King (1965) has suggested that the validity of Wynne-Edwards' theory of population regulation by social mechanisms need not rest entirely upon the unresolved question of group selection; certainly this concept of population regulation warrants careful evaluation regardless of the evolutionary mechanism proposed (it would indeed have been unfortunate, for example, if Darwin's evolutionary theory had been discarded simply because the mechanism of inheritance he proposed was largely incorrect). However, in view of the central role group selection plays in Wynne-Edwards' elaboration of his hypothesis, and the controversy it has aroused, it is necessary to attempt to come to grips with this problem. It is my intention in this paper to dissect and evaluate Wynne-Edwards' interpretation of the process of group selection and to determine the nature of the role such selection may play in the evolution of the social mechanisms he discusses. Perhaps this may mitigate the controversy somewhat and permit clearer thinking on the regulatory mechanisms themselves.

## Wynne-Edwards' Concept of Group Selection

At the outset, Wynne-Edwards fragments natural selection into several categories which he seems to consider functionally as well as conceptually distinct. Selection at the species level, for example, operates where the interests of species overlap and conflict, while individual selection discriminates in favor of individuals that "are better adapted and consequently leave more surviving progeny than their fellows" (1962, p. 18). The greatest attention, however, is given to group selection, since Wynne-Edwards insists that this is the only level at which selection for social behavior can occur. Wynne-Edwards' group selection operates through success or failure of entire groups of individuals; it is, simply stated, natural selection operating at the group level (1962, p. 275). Those groups whose social organization or homeostatic adaptations prove inadequate or unbalanced dwindle away or are forced to overexploit their habitat and thus disappear. More successful groups eventually replace them (1962, pp. 141, 144, 225; 1963, p. 623).

One of the most important premises of this sort of group selection is that animal populations or groups be strongly localized and persistent on the same ground. They are thus largely of common descent, self-perpetuating, and potentially immortal (1962, p. 141; 1963, p. 624). With intergroup gene-flow minimized, the effects of "ordinary" natural selection among individuals can accumulate, adapting them to local conditions; more importantly, such localized group persistence allows group social organization and behavior to develop. Wynne-Edwards, in fact, attributes the evolution of several phenomena, including traditional display sites (e.g., leks), roosting sites, and migratory homing, to their adaptive value in enhancing group localization (1962, pp. 298, 461).

Groups of this sort often have characteristics which are lacking in individuals. These may be social behaviors, such as dominance hierarchies, which cannot exist in a single individual, or behavioral systems in which individual welfare is apparently sacrificed in the interests of the group (so-called altruistic traits). Neither, in the view of Wynne-Edwards, could have evolved by means other than through selection of entire groups, since they are, in fact, group traits. But he clearly believes (1962, p. 141) that the interests of the group are different from those of individuals; what is important in evolution is group survival, and "what is actually passed from parent to offspring is the mechanism for responding correctly in the interests of the group in a wide range of circumstances" (1962, p. 144). Individual selection results in short-term, immediate adaptation to local conditions; group selection is involved in long-term population fitness, which depends on "something over and above the heritable basis that determines the success as individuals of a continuing stream of independent members" (1963, p. 624). Wynne-Edwards, then, views individual selection and group selection as working at cross-purposes, moving toward different adaptive peaks.

Inherent in this concept is a conflict between these selective forces. Social conventions have evolved through group selection, according to Wynne-Edwards, to "safeguard the general welfare and survival of the society, especially against the antisocial, subversive self-advancement of the individual," to shield environmental resources against uncontrolled individuals (1962, pp. 131–132). Individual interests, where they conflict with group welfare, are persistently "overridden" by group selection in the interests of the survival of the group (1963, p. 623).

Wynne-Edwards' visualization of the role of individual selection may be largely responsible for his insistence on separate selective forces. To him, individual selection seems always to imply maximization of individual fecundity — indeed, the chief obstacle he sees to accepting the principle of group selection is the assumption, "still rather widely made, that under natural selection there can be no alternative to promoting the fecundity of the individual, provided this results in his leaving a larger contribution of surviving progeny to posterity" (1963, p. 624). Clearly, Wynne-Edwards believes he has found an alternative in group selection. His image of individual selection stems from Lack's statement (1954, p. 22) that "if one type of individual lays more eggs than another and the difference is hereditary, then the more fecund type must come to predominate over the other (even if there is overpopulation) — unless for some reason the individuals laying more eggs leave fewer, not more, eventual descendents." To Wynne-Edwards, Lack is arguing that selection acts only "at the level of the individual, relentlessly discriminating against those that leave fewer than the maximum possible number of progeny to posterity." To him, it leaves entirely out of account "the overriding effect of group selection, that occurs between one population or society and another, and normally results in fixing the optimum breeding rate for the population as a whole" (1962, p. 485). This conflict of viewpoints is fundamental, and we shall return to it shortly.

It may be well briefly to summarize the essential points of Wynne-Edwards' concept of group selection:

1. Populations or groups have characteristics of their own which are lacking in individuals — these could only have evolved through group selection.

2. Group selection is functionally and conceptually distinct from natural selection at other levels, particularly that of the individual.

3. Often the interests of the group conflict with those of individuals; when this is so, group selection overrides selection at the individual level.

4. Group selection operates through the success or failure of entire groups.

5. These groups are localized and persistent through time, reducing intergroup gene flow.


## EVALUATION OF GROUP SELECTION

Group selection, under a variety of names, has been considered as contributing to the evolution of a number of population phenomena (e.g., Dunbar, 1960; Emerson, 1958, 1960; Sturtevant, 1938; Li, 1955; Thompson, 1958; Lewontin, 1965). Most of these analyses are founded upon the theoretical model of population structure initially elaborated by Wright, and, indeed, Wright's work appears to be the basis of Wynne-Edwards' concept of group selection. Wright has proposed (1931, 1940, 1949, 1951, 1955, and personal communication) that evolution should occur most rapidly and effectively within large populations which are subdivided into small, partially isolated, local populations or demes. In these local "centers" within a population certain gene combinations may arise. These combinations, through pleiotropic interactions, may form an adaptive unit, and confer selective advantages which individual genes could not. If these advantages are of general value to the species they may spread concentrically from their local centers through emigration and increase in frequency in the gene pool of the species. Through an interplay of stochastic and deterministic forces, then, these local centers or demes may become genetically differentiated, increasing the store of genetic variability in the total population above that expected in a panmictic population of the same size.

It is apparent that Wynne-Edwards' concept of group selection parallels Wright's theoretical model. Both stress the role in evolution of selection operating between local populations, but there is an important difference in emphasis. Wright views the individual as "the direct object of selection" (1955, p. 20); selection between populations only *supplements* individual selection within them (*ibid.*, p. 17). Group selection can do no more than "correct in part the shortcomings" of individual selection from the standpoint of the species as a whole (*ibid.*, p. 21), through the spread of gene combinations of adaptive value. Group selection in Wright's view, then, may be important in the evolution or divergence of populations, but it is founded upon and only supplements individual selection and adaptation. The balance and interplay of Wright's selective forces contrast sharply with the independence and overriding power of Wynne-Edwards' group selection.

There also appears to be disagreement on the amount and significance of gene flow occurring between demes. Wright holds that some immigration from adjacent demes may increase the genetic variability of a deme and prevent fixation, and that adaptations of general value to the species may spread through emigration or outbreeding. Wynne-Edwards, on the other hand, concedes only a "trickle" of interpopulation gene flow, and regards localized persistence of groups and minimal intergroup gene flow as prerequisites of group selection (1962, p. 141). Elsewhere (*ibid.*, pp. 245, 355), however, he observes that some forms of dispersal may promote a "desirable amount of outbreeding." As used by Wynne-Edwards, group selection also appears to operate in an "all or none" manner, entire groups either persisting or dwindling away. Not only does this imply an unrealistic lack of variability or flexibility in group gene pools, but it also disallows the spread by emigration of local gene combinations which to Wright is group selection (1940, p. 239; personal communication).

Wright's theoretical model, while generally accepted in principle, has been criticized on several counts. Early criticisms by Fisher and Ford (1947) have been answered (Wright, 1948), but more recently Simpson (1953, p. 123) and Mayr (1963, p. 520) have pointed out that the degree of interdeme isolation required by the model is probably rarely achieved in natural situations. The criticism revolves about the unresolved question of the magnitude at which gene flow effectively disrupts local divergence.

Wynne-Edwards' insistence upon the necessity of group selection to his hypothesis and its pervasive overriding power appear to stem in part from his inability to explain in any other way the evolution of characteristics which transcend individual traits. Surely, many socially organized groups do have attributes which are absent in a random assemblage of individuals, but, as Dobzhansky (1955, p. 14) has pointed out, "an individual is not identical to the whole to which it belongs; but a population as a whole is what it is only because of the individuals of which it is composed." Populations are of course composed of individual members, and population changes are essentially a matter of changes in the individuals comprising the population (Milne, 1957, p. 261); what is always selected are individuals (Birch, 1960, p. 13; Dobzhansky, 1955, p. 12; Lewontin, 1965, p. 302; Williams, 1966). Conversely, however, individuals do not exist in nature completely isolated from other individuals — they are members of populations with varying degrees of integration and cohesion, and no matter what the extent of integration, other members of the population exert effects upon each individual. The population, in other words, is itself an environmental factor for the individual (Milne, 1957; Andrewartha and Birch, 1954; Dobzhansky, 1957; Nicholson, 1957). As such, it is part of the total environmental complex to which individuals are adapted, and it may often constitute a selective force of considerable magnitude. Thus it must inevitably be adaptive for individuals to react efficiently in accord with other members of the population. As a consequence of these coadaptive responses of individuals, the population itself may develop a certain degree of integration and social organization which transcends that of individual members. Such "group attributes," however, are firmly founded upon individual responses. In other words, behaviors which conform to the "interests" of the group, in Wynne-Edwards' sense, are basically the result of individual adaptation to individual "interests." Wynne-Edwards' separation of these interests may result from a neglect or misunderstanding of the importance of the role of the population in the complex of environmental selective forces impinging upon the individual.

Other difficulties exist in Wynne-Edwards' interpretation of the evolution of social behaviors: as Maynard Smith (1964) has pointed out, it does not explain how a behavior initially arises within a group. Group selection, even as Wynne-Edwards uses it, can only account for the persistence and spread of entire groups, all (or nearly all) members of which possess the appropriate behavior. Considered at the level of the individual, however, this problem vanishes, since it seems apparent that any behavior which increases even slightly the over-all adaptiveness or efficiency of an individual in relation to its fellows will be favored by selection, and eventually spread through the population by the slight edge in differential reproduction it confers. Also, group selection would appear to be too slow to function effectively in evolution — what corresponds to the life duration in individual selection must be more like the duration of the existence of the group in group selection. Since the number of groups in a population or a species is far less than the number of individuals, it would also seem that the selectable variance available in group selection must be much less than in selection at the individual level (Fisher, 1958, p. 50).

But what of the apparent paradox of a behavior, evolved through differential survival and fertility of individuals, which seems to lead to the sacrifice of some members of the population? Such "altruistic" traits (e.g., alarm calls; distraction displays; various aspects of parental care, etc.) have often been cited as phenomena which can only be explained through the action of group or "kin" selection (Wynne-Edwards, 1962; Maynard Smith, 1964, 1965; Hamilton, 1963; Haldane, 1955). Wynne-Edwards, of course, sees no problem in explaining altruistic behavior, since it benefits the group and is thereby preserved by group selection. Hamilton rejects group selection because he feels it is too slow. To Maynard Smith the difficulty in group selection is explaining how altruistic traits might initially spread through a population: if by natural selection they presumably could spread by individual processes; genetic drift, he believes, is the only way such characteristics could spread to all members of a group by group selection, and he holds that the conditions required for this to occur are too rarely realized in nature (it might be noted that Maynard Smith's logic in this area is not entirely convincing: why, for example, is a group "more likely to split into two" if "all members of a group require some characteristic which, although individually disadvantageous, increases the fitness of the group?" — 1964, p. 1145). Both Hamilton and Maynard Smith resort to explanations based on "kin selection." In kin selection the selectively favored individuals are close relatives (progeny, siblings, etc.) of the "altruistic" individual. If relatives live together in family groups there will be more opportunities for kin selection to be effective, particularly if the population is divided into partially isolated groups, but such subdivision is not essential. The effectiveness of kin selection is directly proportional to the degree of relationship of the individuals concerned, and altruistic behavior which benefits individuals regardless of relationship may evolve only if the disadvantage involved is very slight and the average neighbor is not too distantly related (Hamilton, 1963).

There are two points which may clarify the paradox of altruistic behavior somewhat. In the first place, kin selection would seem to be no different, functionally, from individual selection. Individual selection operates through differential reproduction, genotypes leaving the most nearly optimal number of adapted progeny being, in the long run, selectively favored (Mayr, 1963, p. 183). Any genetically determined behavior of an individual which tends to enhance the chances for survival of its progeny will be advantageous, so long as it is not outweighed by disadvantages in other contexts. The individuals selected will be those whose parents demonstrated the most adaptive behavior (as Birch, 1960, p. 13, has pointed out in another context). Thus, the evolution of many "altruistic" behaviors may readily be explained by selection at the individual level. Secondly, I think the label "altruistic" should be applied to behavior patterns with extreme caution. It seems to have been rather tacitly assumed that altruistic traits are disadvantageous or detrimental to the individual displaying them, when in fact they may accord some unseen advantages. "Injury-feigning" birds, for example, are rarely if ever caught, and alarm calls usually have sonic characteristics which make the source of the call difficult to locate (Marler, 1957). Other examples of seemingly altruistic behavior may be complexes of characters which together appear to be disadvantageous, but which separately are extremely adaptive responses to different selective forces (Tinbergen, 1965). Many "altruistic" traits, then, may be comparable to morphological features which were considered "nonadaptive" until detailed investigations revealed very real adaptive functions (Cain, 1964).

Previously, mention was made here of Wynne-Edwards' view that individual selection implies to many workers an absolute maximization of fecundity, with the consequence that submaximal reproduction can only be explained by group selection. But it appears that he has missed an essential part of what is generally understood to constitute individual selection.

Surely what counts in evolution is indeed the contribution made by genotypes to the gene pool of following generations (Mayr, 1963). But, contrary to the assumption made by Williams (1966), this contribution would generally seem to be made through the most adaptive manner and rate of reproduction in a given environment, and *not* simply by maximizing reproduction regardless of long-term effects. Selection does not favor a high reproductive rate per se, but a reproductive rate profitable from the standpoint of producing the optimal *eventual* number of successful progeny (Pimentel, 1961, p. 74; Lack, 1954, p. 22; Amadon, 1964; Perrins, 1965). Birch (1960, p. 10) has pointed out that natural selection will tend to maximize the potential rate of increase of a species *for the environment in which it lives*, since any genetic factors, or combinations of factors, which increase the chance of genotypes that possess them contributing more individuals to following generations will be maintained by selection. This, of course, does not mean that those types presently increasing will indefinitely continue to hold the edge in differential reproduction, or that present rates of increase will always be the most adaptive; in a fluctuating environment selective forces change through time, so that present rates of increase may not necessarily be good indicators of which type will contribute most to remote future generations (R. H. MacArthur, personal letter, 1966). Nor does it necessarily imply that birth rate must increase or death rate decrease, but only that the difference between the two be maximized. Actually, it may be better to speak of "optimization" of reproductive rate rather than "maximization": natural selection operates in favor of those individuals who leave, in the most efficient manner, and in the long run, the optimal number of offspring (i.e., neither too many nor too few) in relation to the environmental forces influencing reproduction (including the long-term carrying capacity of the environment). Obviously, if the environmental forces fluctuate during the time interval under consideration, the optimal reproduction rate will also change.

Thus it seems that here, as elsewhere, reproductive phenomena can be explained by "ordinary" natural selection operating at the level of the individual, without recourse to group selection processes.

## SELECTION AND REGULATORY BEHAVIORS

It remains to examine briefly how some of the forms of social behavior discussed by Wynne-Edwards may be explained without resorting to group selection.

*Territorial Behavior.* Wynne-Edwards (1962, 1965) proposes that territorial competition is one of a number of forms of "property tenure" which regulate population size and insure an adequate supply of resources (usually food) for their holders. To Wynne-Edwards this regulation is the sole function of territorial systems, but, while there is evidence that territory does indeed serve such a function on occasion (e.g., Tompa, 1962; Carrick, 1963; Gibb, 1956; Orians, 1961), approaches stressing convergence and diversity of function (e.g., Brown, 1964; Crook, 1965; McBride, 1964; Tinbergen, 1957) must also be considered. Regardless of the function it serves, the evolution of territorialism can be adequately explained by individual selection, as Orians and Wilson (1964), Maynard Smith (1964), Brown (1964), and Crook (1965) have indicated. If individuals which establish territories but are too aggressive raise fewer offspring (as visualized in the concept of aggressive neglect; see Ripley, 1961), while excessively timid individuals fail to establish a territory, then the individuals with the optimal balance of genetic factors regulating territorial aggressiveness will be expected to leave more surviving and reproducing offspring. In other words, the type and degree of territorial aggressiveness exhibited by these individuals will in time become the new population norm (Brown, 1964; Maynard Smith, 1964). In the same manner, the size and type of territory optimal for any species will be established through selection of individuals which hold the most efficient territories with respect to the function they serve and their ecological context. Again, Wynne-Edwards' argument appears to fail "primarily because it does not take account of the fact that changes in gene frequency are the result of competitive advantages accruing to individual genotypes rather than to the group as a whole" (Brown, 1964, p. 167).

*Dominance Hierarchies.* Wynne-Edwards uses the hierarchy as an example of a typical group character, manifested in a collective group but absent in individuals, and considers the "conventional processes" by which hierarchies are established to be group characteristics as well. Hierarchies are viewed as playing an important role in regulating populations, since the "surplus tail" of the hierarchy may either be disposed of or retained according to the availability of resources. The function of the hierarchy is therefore "always to identify the surplus individuals whenever the population-density requires to be thinned out" (1962, p. 139). Subordinate individuals may lose their sole chance of reproduction by passively yielding to dominant individuals, but the survival of the group as a whole depends upon their compliance to the hierarchy (1963, p. 625). Wynne-Edwards, of course, proposes that hierarchies have arisen through group selection, but curiously he attributes the development of symbolic characters or ornaments (horns, plumes, etc.) which contribute to determining social status to selection operating on the individual (1962, p. 139). Elsewhere (*ibid.*, p. 247), he observes that dominant individuals generally show only average development of such characters, and that the most extravagant ornaments do not necessarily confer the highest status. Dominance is based, of course, on a complex of features, and the ingredients may vary from one individual to the next. It is not necessary to assume, as does Wynne-Edwards (1963, p. 626), that because the characteristics that confer dominance are diverse, selection for dominance at the individual level must be dissipated through the gene pool of the population and thus be ineffective. If a variety of features can combine in various ways to confer dominance, then, through the reproductive advantage which often goes with dominance status, this variety will be passed on to subsequent generations by simple individual selection. In addition, hierarchies depend upon the ability of individuals to identify rapidly other individuals in the hierarchy.

Thus, the relatively wide range of variation in development of symbolic structures noted by Wynne-Edwards may serve to facilitate individual recognition rather than affect social status per se.

The regulation of breeding through dominance status can, as Wynne-Edwards observes, effect a socially mediated selection at the individual level. But it would seem that the hierarchical system may itself be a product of individual selection as well, since, by stabilizing the aggressive relationships between individuals and limiting active competition to the formative phases of the hierarchy, it may lead to a more efficient energy expenditure in terms of individual reproduction. Since the features that confer dominance are not fixed, any subordinate individual, although initially excluded from breeding, may later attain a position of relative dominance, and, in the long run, exceed the reproductive potential expected had it not complied with the system. It is well known, for example, that in many species age generally confers higher social status. It may therefore be individually advantageous to submit and live to breed later than to persist in fighting and leave few, if any offspring. Also, since the population is an important feature of the environment to which any individual becomes adapted, hierarchical systems (or, indeed, any form of social organization), by increasing the stability of aggressive relationships within the population, may exert a feedback on individual adaptation. It would seem, therefore, that such social behaviors may be individually adaptive in themselves, and not simply "incidental statistical by-products" of other individual adaptations, as Williams (1966, pp. 218, 240) has suggested.

*Societies in Primitive Man.* Wynne-Edwards devotes considerable attention to primitive human societies, drawing extensively from the work of Carr-Saunders (1922). Localization of human groups and the establishment of tribal territories are pictured as leading to the formation of partially isolated groups that are "exactly comparable with those we find elsewhere in the animal world" (1962, p. 188), and between which group selection may operate. Indeed, primitive human societies, with their diverse forms of property tenure, traditions, moral codes, laws, hierarchies of social status, and so on, appear to Wynne-Edwards to offer some of the clearest examples of the action of group selection. He notes (1962, p. 249) that critics of group selection "must face yet another paradox here, namely, how selection at the individual level could lead to the development of characters, such as those of venerability and the prolongation of old age, which do not take effect until after the period of procreation is past." In terms of group selection, on the other hand, he feels there is no difficulty in "understanding the ascendancy of those human groups that are best able to benefit by the councils of their elder statesmen." But there is in fact no contradiction here. Cultural transmission and tradition play a far greater role in human societies than in those of any other animals, and their evolution is probably influenced to a far greater degree by cultural than by organic factors (Huxley, 1960; Hallowell, 1960). Cultural transmission may indeed induce group isolation and permit the accumulation through time of cultural features distinctive to any particular group, but this isolation is, of course, of quite a different nature from that found in other animal species. Actually, there is evidence that primitive human societies were not nearly as isolated as Wynne-Edwards supposes. Many groups had detailed knowledge of neighboring cultures, and often intergroup communication and influence were extensive, as between the Malayan regions and the Australian aborigines (Warner, 1932). In addition, prolongation of old age and retention of "senile" individuals in primitive human societies are by no means common, but where they do occur they may be the product of kin selection rather than group selection (i.e., post-breeding individuals who maintain a role in the family, such as a baby-sitting grandmother, are effectively increasing the chances for survival of their descendants). The virtual absence of such phenomena in other animal groupings indicates, however, that they are a product of the "cultural evolution" which characterizes the evolution of human societies.

## CONCLUSIONS

I think it is clear that many of the social phenomena Wynne-Edwards discusses may be explained without recourse to group selection, at least in the sense Wynne-Edwards has used it. While some form of group or population selection probably does play a role in speciation and evolution, Wynne-Edwards has emphasized its role to the point of considering it as replacing selection at the individual level. But the individual must be, as we have seen, the basic unit of selection, and many of the social attributes analyzed by Wynne-Edwards have no doubt evolved at that level. He has, in effect, missed the very important point made by Wright — that selective forces are interwoven, that group or interdeme selection may only supplement, not override, individual selection. Neither does there seem to be any realization that the population is a very important component of the complex of environmental forces impinging on the individual, and that through positive feedback this interrelation of population and individual may lead to high levels of social organization by "ordinary" natural selection.

Despite these shortcomings, however, the theory of population regulation through social mechanisms advanced by Wynne-Edwards is certainly worthy of much serious study, and it should not be discarded simply because it appears to rely on group selection. As Kalela (1957) has observed, there is no reason to suppose that theories of population regulation must be mutually exclusive, that different mechanisms may not be regulating populations, to different degrees, in different cases. Where Wynne-Edwards' theory seems to apply it may be quite useful: where it obviously does not apply it may be discarded, but in all studies of population phenomena it should at least be considered. If there is to be continuing controversy, then, it should center on the relative merits and the applicability of the hypotheses explaining population regulation, and not on the evolutionary mechanisms involved.

## Acknowledgments

Many people have been instrumental in the genesis and development of the ideas expressed in this paper. Roger Evans, James Drescher, and others of the "fourth floor group" in Birge Hall at the University of Wisconsin provided many hours of stimulating discussion. Conversations with Drs. James F. Crow, Sewall Wright, and members of their population genetics seminar were tremendously helpful. Drs. James F. Crow, John T. Emlen, Masakazu Konishi, William G. Reeder, and Sewall Wright read drafts of the manuscript and offered numerous suggestions. In the end, of course, Prof. V. C. Wynne-Edwards, through his broad and thoughtful book, has provided the greatest stimulation. Any errors in interpretation or fact must, of course, be my own.

REFERENCES

Amadon, D. 1964. The evolution of low reproductive rates in birds. *Evolution*, 18, 105–110.
Andrewartha, H. G., and L. C. Birch. 1954. The distribution and abundance of animals. Univ. Chicago Press, Chicago, 782 pp.
Birch, L. C. 1960. The genetic factor in population ecology. *Amer. Nat.*, 94, 5–24.
Braestrup, F. W. 1963a. Special review of "Animal Dispersion in Relation to Social Behavior." *Oikos*, 14, 113–120.
———. 1963b. The functions of communal displays. *Dansk. Ornith. Foren. Tidsskr.*, 57, 133–142.
Brown, J. L. 1964. The evolution of diversity in avian territorial systems. *Wilson Bull.*, 76, 160–169.
Cain, A. J. 1964. The perfection of animals. In Carthy, J. D., and C. L. Duddington (eds.), *Viewpoints in Biology*, 3, 36–63.
Carrick, R. 1963. Ecological significance of territory in the Australian Magpie, *Gymnorhina tibicen. Proc. 12th Int. Ornith. Congr.*, pp. 740–753.
Carr-Saunders, A. M. 1922. *The population problem: a study in human evolution.* Oxford Univ. Press, Oxford.
Crook, J. H. 1965. The adaptive significance of avian social organizations. *Symp. Zool. Soc. Lond.*, 14, 181–218.
Dobzhansky, T. 1955. A review of some fundamental concepts and problems of population genetics. *Cold Spr. Harb. Symp. Quant. Biol.*, 20, 1–15.
———. 1957. Mendelian populations as genetic systems. *Cold Spr. Harb. Symp. Quant. Biol.*, 22, 385–393.

Dunbar, M. J. 1960. The evolution of stability in marine environments. Natural selection at the level of the ecosystem. *Amer. Nat.*, 94, 129–136.

Elton, C. S. 1963. Self-regulation of animal populations. *Nature*, 197, 634.

Emerson, A. E. 1958. The evolution of behavior among social insects. In Roe, A., and G. G. Simpson (eds.), Behavior and evolution. Yale Univ. Press, New Haven, pp. 311–335.

——. 1960. The evolution of adaptation in population systems. In Tax, S. (ed.), The evolution of life. Univ. Chicago Press, Chicago, pp. 307–348.

Fisher, R. A. 1958. The genetical theory of natural selection. Dover, New York, 291 pp.

——, and E. B. Ford. 1947. The spread of a gene in natural conditions in a colony of the moth *Panaxia dominula* L. *Heredity*, 1, 143–174.

Gibb, J. 1956. Territory in the genus *Parus*. *Ibis*, 98, 420–429.

Haldane, J. B. S. 1955. Population genetics. *New Biology*, 18, 34–51.

Hallowell, A. I. 1960. Self, society, and culture in phylogenetic perspective. In Tax, S. (ed.), The evolution of man. Univ. Chicago Press, Chicago, pp. 309–371.

Hamilton, W. D. 1963. The evolution of altruistic behavior. *Amer. Nat.*, 97, 354–356.

Huxley, J. 1960. The emergence of Darwinism. In Tax, S. (ed.), The evolution of life. Univ. Chicago Press, Chicago, pp. 1–21.

Kalela, O. 1957. Regulation of reproduction rate in subarctic populations of the vole Clethrionomys rufocanus (Sund.). *Ann. Acad. Scien. Fenn.*, Series A, IV, No. 34, 60 pp.

King, J. A. 1965. Social behavior and population homeostasis. *Ecology*, 46, 210–211.

Koskimies, J. 1955. Ultimate causes of cyclic fluctuations in numbers in animal populations. Papers on Game Research (Helsinki), No. 15, 29 pp.

Lack, D. 1954. The natural regulation of animal numbers. Oxford Univ. Press, Oxford, 343 pp.

——. 1965. Evolutionary ecology. *J. Anim. Ecol.*, 34, 223–231.

Lewontin, R. C. 1965. Selection in and of populations. In Moore, J. A. (ed.), Ideas in modern biology. Natural Hist. Press, Garden City, pp. 299–311.

Li, C. C. 1955. Population genetics. Univ. Chicago Press, Chicago, 366 pp.

Marler, P. 1957. Specific distinctiveness in the communication signals of birds. *Behaviour*, 11, 13–39.

Maynard Smith, J. 1964. Group selection and kin selection. *Nature*, 201, 1145–1147.

——. 1965. The evolution of alarm calls. *Amer. Nat.*, 99, 59–63.

Mayr, E. 1963. Animal species and evolution. Harvard Univ. Press, Cambridge, 797 pp.

McBride, G. 1964. A general theory of social organization and behaviour. *Univ. Queensland Pap. Dept. Veterinary Sci.*, 1, 75–110.

Milne, A. 1957. Theories of natural control of insect populations. *Cold Spr. Harb. Symp. Quant. Biol.*, 22, 253–267.

Nicholson, A. J. 1957. The self-adjustment of populations to change. *Cold Spr. Harb. Symp. Quant. Biol.*, 22, 153–172.

Orians, G. H. 1961. The ecology of blackbird (*Agelaius*) social systems. *Ecol. Monogr.*, 31, 285–312.

——, and M. F. Willson. 1964. Interspecific territories of birds. *Ecology*, 45, 736–745.

Perrins, C. M. 1965. Population fluctuations and clutch-size in the Great Tit, *Parus major* L. *J. Anim. Ecol.*, 34, 601–647.

Pimentel, D. 1961. Animal population regulation by the genetic feed-back mechanism. *Amer. Nat.*, 95, 65–79.

Ripley, S. D. 1961. Aggressive neglect as a factor in interspecific competition in birds. *Auk*, 78, 366–371.

Selander, R. K. 1965. On mating systems and sexual selection. *Amer. Nat.*, 99, 129–141.

Simpson, G. G. 1953. The major features of evolution. Columbia Univ. Press, New York, 434 pp.

Sturtevant, A. H. 1938. Essays on evolution. II. On the effects of selection on social insects. *Quart. Rev. Biol.*, 13, 74–76.

Thompson, W. R. 1958. Social behavior. In Roe, A., and G. G. Simpson (eds.), Behavior and evolution. Yale Univ. Press, New Haven, pp. 291–310.

Tinbergen, N. 1957. The functions of territory. *Bird Study*, 4, 14–27.

———. 1965. Behavior and natural selection. In Moore, J. A. (ed.), Ideas in modern biology. Natural Hist. Press, Garden City, pp. 521–542.

Tompa, F. S. 1962. Territorial behavior: the main controlling factor in a local song sparrow population. *Auk.*, 79, 687–697.

Warner, W. L. 1932. Malay influences on the aboriginal cultures of northeastern Arnhem Land. *Oceania*, 2, 476–495.

Williams, G. C. 1966. Adaptation and natural selection. Princeton Univ. Press, Princeton, N.J., 307 pp.

Wright, S. 1931. Evolution in Mendelian populations. *Genetics*, 16, 97–159.

———. 1940. Breeding structure of populations in relation to speciation. *Amer. Nat.*, 79, 232–248.

———. 1948. On the roles of directed and random changes in gene frequency in the genetics of populations. *Evolution*, 2, 279–295.

———. 1949. Adaptation and selection. In Jepson, G. L., G. G. Simpson, and E. Mayr (eds.), Genetics, paleontology and evolution. Princeton Univ. Press, pp. 365–389.

———. 1951. The genetical structure of populations. *Ann. Eugenics*, 25, 323–354.

———. 1955. Classification of the factors of evolution. *Cold Spr. Harb. Symp. Quant. Biol.*, 20, 16–24.

Wynne-Edwards, V. C. 1962. Animal dispersion in relation to social behavior. Hafner, New York, 653 pp.

———. 1963. Intergroup selection in the evolution of social systems. *Nature*, 200, 623–626.

———. 1965. Social organization as a population regulator. *Symp. Zool. Soc. Lond.*, 14, 173–178.

# The Natural Selection of Self-Regulatory Behavior in Animal Populations

## DENNIS CHITTY

The object of this paper is to discuss ways of testing the hypothesis that all species of animals have a form of behavior that can prevent unlimited increase in population density (Chitty, 1960, 1964, 1965). "Behavior" is taken to include the numerous manifestations of hostility discussed by Tinbergen (1957), and the first assumption is that all species have some form of dispersion mechanism, or method of spacing themselves out by avoiding, threatening, or otherwise influencing certain other members of the same species. The pattern of dispersion is assumed to depend partly on the properties of the individuals (genotype, age, experience, hormone balance, etc.), and partly on the properties of their environment (weather; amount, kind, and distribution of food and cover; number and kind of other animals present, etc.). The second assumption is that this

From *Proceedings of the Ecological Society of Australia,* 2 (1967), 51–78.

behavior persists only because it has survival value for the individual and is constantly being selected for.

While assuming that this behavior occurs in all *species*, the hypothesis does not imply that all *populations* are thereby self-regulated: merely that within any species one can expect to find a class of population whose rate of increase is slowed down and finally stopped because of the way its members interact with one another. The hypothesis also recognizes that this supposed self-regulatory mechanism is not always effective in preventing populations from reaching outbreak proportions, such as those sometimes observed in rodents and defoliating insects; the problem is to explain why such outbreaks do not occur more often. Thus no claim is made that the mechanism is either necessary to prevent the increase of all populations, or sufficient to prevent others from depleting their food supply; we cannot, in fact, make any statements that can be tested on all populations of all species. Yet a *sine qua non* for any scientific hypothesis is that it be stated in a way that can be tested, which means that some form of generalization is required. Testing consists of trying to falsify this generalization, since evidence in favor of a hypothesis is generally useless except when it results from the failure of such attempts (Cohen and Nagel, 1934; Medawar, 1957; Popper, 1959, 1963; Platt, 1964).

Generalizations, then, do not necessarily assert the truth, but become part of a logical device for discovering falsity. The more widely a hypothesis can be tested, the greater the chance of evidence being found that not only contradicts it but explains its failure. Most of the evidence for the present hypothesis comes from tests of this sort carried out on small mammals; but really crucial evidence is lacking, since neither individual behavior nor population genetics has so far received much attention from those working on mammal populations. It is thus fortunate that the assumptions on which the hypothesis depends are completely general; hence, contradictory evidence, regardless of the species from which it is obtained, should enable one to test the present interpretation: it cannot both be true for the relevant populations of small mammals and false for those of other groups. This being so, it is clearly advisable to study those species best suited for an experimental attack on the problem. An attempt to enlist the help of entomologists has already been made (Chitty, 1965); as Klomp (1964) has shown, self-regulation occurs among the insects.

The argument in the following pages is set forth as follows: (a) many small mammal populations have a recognizably distinct form of fluctuation, which is not explained by conventional hypotheses; (b) these fluctuations are consistent with the idea that there are changes with population density in the survival value of certain kinds of behavior; (c) this and all other hypotheses must be tested by methods designed to refute them and thus to avoid mere confirmation of *a priori* ideas; (d) the testing of hypotheses about the more stationary populations is complicated by changes in numbers due to changes in irrelevant variables, and by lack of adequate comparative date between regulated and unregulated populations.

## THE PHENOMENON IN CERTAIN MAMMAL POPULATIONS

The present ideas have developed from an attempt to explain the recurrent declines in numbers in populations of the Field Vole, *Microtus agrestis* (Elton, 1931, 1942). Lack of success in finding a solution to this problem is partly due to the difficulties illustrated in Table 1, which is based on an early study (Chitty, 1952). As shown, a decline to scarcity occurred after the second of two winters of abundance on area A, but not after the first. Only 10 per cent instead of 40 per cent of the animals survived this second winter, and they did not grow to full size; yet survival and growth were good at the same time on other areas. Clearly, then, high numbers in autumn and winter are not a sufficient condition for a decline; indeed it can occur among overwintered populations that are relatively sparse (Chitty and Chitty, 1962a). Since Hamilton (1937, 1941) had already observed this pattern of change, I described it as the H type of decline (Chitty, 1955), to distinguish it from the G type (after Godfrey, 1955), which

Table 1 : *Changes in Numbers and Body Weight of* Microtus Agrestis *(from Chitty, 1952)*

|  |  | 1936 | 1937 | | 1938 | |
|---|---|---|---|---|---|---|
|  |  | Sept | May | Sept | May | Sept |
| *Area A* |  |  |  |  |  |  |
| Approx. no./acre |  | 300 | 120 | 300 | 30 | 1? |
| Mean body wt (g.) ± s.e. | males | | 29.06 ± 0.42 n = 62 | | 24.04 ± 0.48 n = 53 | |
| | females | | 24.07 ± 0.46 n = 44 | | 18.42 ± 0.33 n = 46 | |
| *Other areas* |  |  |  |  |  |  |
| Abundance |  | — | — | — | high | high |
| Mean body wt (g.) ± s.e. | males | | — | | 30.45 ± 0.81 n = 51 | |
| | females | | — | | 26.19 ± 0.73 n = 34 | |

continues unchecked throughout the breeding season. Krebs (1964b) observed both patterns among lemmings (Figure 1), and has described others in populations of *Microtus californicus* (Krebs, 1966). High rates of loss also occur among relatively sparse populations of Snowshoe Hares (Figure 2). Although there are many differences between populations, yet where enough data are available they suggest that a common class of events is involved. Not only are these declines recurrent phenomena, but they are attended by characteristic changes in the distribution of body weights and survival rates. Thus one advantage of studying these declines is that they are easily recognizable instances of the phenomenon we want to explain. In addition they occur fairly often, are sufficiently pronounced to be relatively unaffected by irrelevant environmental variables, and may occur when other

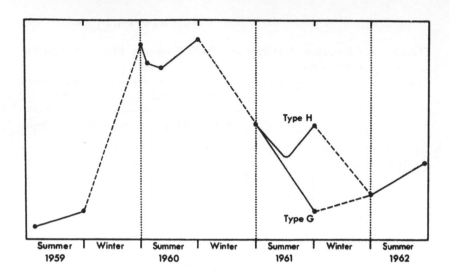

FIGURE 1: Generalized changes in population density of the lemmings *Lemmus trimucronatus* and *Dicrostonyx groenlandicus*; log scale. (From Krebs, 1964b.)

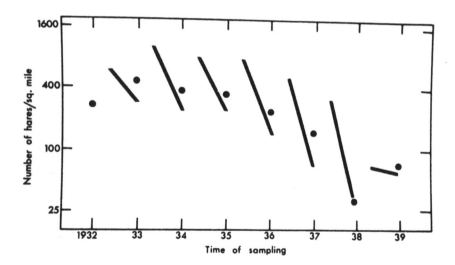

FIGURE 2: Population density of Snowshoe Hares (*Lepus americanus*) in February 1932–9 ( • ). Sloping lines show the difference between estimates of the numbers born and numbers alive the following February. (From Chitty, 1964, based on Green and Evans, 1940.)

populations are increasing or stationary, thus permitting one to obtain properly comparative data (Table 1; also Figures 4 and 5 discussed below).

Since declines of this type are uncommon elsewhere in the animal kingdom it is fair to ask whether their study has any general significance. The assumption made here is that these particular fluctuations are simply well-marked effects of a general process, whose more familiar effect is to keep numbers fairly stationary, a state sometimes observed even in vole populations (Summerhayes, 1941, Table 5).

One may, of course, make quite different assumptions, provided one tests their implications. Ford (1961) writes: "Considerable attention has in the past been directed to a limited group of special instances in which alternating periods of abundance and rarity coincide over extensive areas. Such cycles are doubtless climatic in origin, and have been associated among other things with the waxing and waning of sun-spots. Yet they have in fact been detected in very few species, a small selection of rodents and animals of value to the fur trade, and they are of negligible importance compared with the automatic fluctuations in numbers to which organisms in general are subject." A slightly updated version is given by Ford (1964).

The present interpretation of this phenomenon differs from that put forward by Chitty (1960). At that time I still supposed that interactions between animals had their main effects on viability, and that declines in the numbers of Field Voles were sufficiently explained by the consequent increase in severity of action of the normal forces of mortality, such as bad weather. In other words I supposed there was a systematic change in the properties of the individuals but not of their environment. This idea was found to be inadequate (Newson and Chitty, 1962), since animals were clearly not in a serious pathological condition just prior to their decline in numbers. It therefore seemed that some environmental component, not previously taken into account, was probably responsible for the changes in survival; and of those not already excluded the most likely was that attributable to the activities of the animals themselves.

Andrewartha and Birch (1954) have written that "The difficulty of thinking of the 'environment of the population' is that it leaves out half the picture." Even this may be an understatement. Early vole studies had shown that the "environment of the population" seemed to be favorable during the declines; and although I had assumed that the "environment of the individual" was unfavorable at the high densities prior to a decline, I had not imagined that this condition would persist at the lower densities during the decline itself, especially during that of type H.

One other point should be made about my 1960 paper, namely that it dealt with the occurrence of qualitative changes and their ecological implications (cf. Wellington, 1960) and not with their explanation in terms of selective advantage. At the time it was difficult to imagine what selective advantage to look for, other than some unknown physiological property that enabled certain individuals to be relatively unaffected by interference (Chitty, 1958). Even so it was difficult to see how selection could act so drastically in such a short number of generations. It is worth emphasizing this difficulty, which has been the main barrier to looking for a solution in terms of population genetics, especially as there is also a great difference in mean adult body weight between a peak population and one that has suffered a severe decline (Chitty, 1952; Chitty and Chitty, 1962b; Newson and Chitty, 1962; Krebs, 1964b, 1966). The animals with the low body weights in Table 1 were either the $F_1$ progeny of the animals of the previous spring, or the backcrosses between the $F_1$ females and the parental males. How then were such big qualitative differences to be accounted for?

The most likely solution came from finding that declining populations consisted of a mixture of animals with potentially good and potentially poor growth rates (as revealed by their performance when taken into the laboratory), and from the suggestion that in the field "some unknown aspect of their own behavior may have been at least partly responsible for the poor survival, growth, and reproduction of the voles" (Newson and Chitty, 1962). From this grew the idea that the animals with the inherently poor growth rates were inhibiting the growth and survival of the others. Thus one could now conceive of large population effects, partly phenotypic and partly genotypic, resulting from the behavior of a few individuals. One highly aggressive animal that could deny living space to many others, or inhibit their growth or reproduction, would clearly have a huge selective advantage. Even so it would have been difficult to entertain this idea had there not, since the earlier days of these studies, been a fundamental change in outlook on the part of the population geneticists. Ford (1964) refers to "the recent recognition that advantageous qualities are frequently favored or balanced in particular environments by far greater selection-pressures than had hitherto been envisaged."

A summary of present ideas has already been given by Krebs (1964b); here we shall refer to two points only. First, the loop in Figure 3, indicating a systematically increasing selection pressure, explains how a declining population might come to contain a high proportion of aggressive animals. The effects of interaction would thus be more severe than those in a higher density population of more docile animals. The increasingly severe loss of young Snowshoe Hares (Figure 2) is also consistent with such selection becoming more pronounced after the initial years of high numbers.

Secondly, this selection is shown as having an adverse effect on viability. Tinbergen (1957) has discussed some of the disadvantages to which highly aggressive animals are subject; and the time lost in fighting or guarding the nest and the difficulties between mates are well illustrated in the Gannet, *Sula bassana* (Nelson, 1964, 1965). Clearly, then, selection cannot go too far in this direction, especially if there are associated disadvantages in attributes other than behavior.

If the "environment of the individual" in stationary or declining populations is indeed different from that of the individual in an expanding population, then it is reasonable to suppose that animals best fitted for the one environment are less well fitted for the other. The generally good reproductive performance of expanding populations (Davis, 1951; Chitty and Chitty, 1962b; Newson, 1963; Krebs, 1964b, 1966) is consistent with this idea,

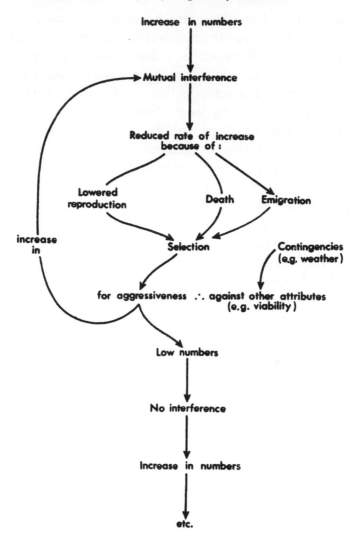

FIGURE 3: Postulated self-regulatory feedback system in small mammal populations. (From Krebs, 1964b.)

but also with many others (Christian and Davis, 1964; Smyth, 1966). Genotypes that are usually eliminated may also survive better during these periods of rapid increase (Ford, 1961), but there is no reason to suppose that they will increase faster than genotypes already adapted to the local hazards.

Despite the evidence against the original belief that declining populations are markedly less viable than normal (Chitty, 1952), there is some theoretical justification for believing that adverse changes in attributes are still worth looking for. Mather (1961) has stated the case as follows:

> selection arising from competition does not necessarily make succeeding generations of individuals of that species more fit to meet circumstances other than those of the competition itself. Selection arising from competition will favor nothing but the ability to survive that particular type of competition. Indeed the selective rise of a special type of competitive power may be accompanied by a reduction in other components of fitness.

The scheme represented by Figure 3 differs from two somewhat similar schemes in at least the following ways: it differs from that of Christian, Lloyd, and Davis (1965), who claim that phenotypic changes alone are adequate to explain why populations decline in numbers; it differs from that of Koskimies (1955); Kalela (1957); Snyder (1961), Brereton (1962), Wynne-Edwards (1962), and others who explain regulation in terms of mechanisms produced by group selection. Differences from other explanations are that (1) population density is assumed to be only one of many environmental variables affecting population trends and not necessarily the best for predicting them, and (2) mortality factors are assumed to be among the local variables about which predictions of rather limited value can be made from one population to another or from one year to the next. But while I attribute no special relevance to the so-called density-dependent mortality factors, and thus agree with Andrewartha and Birch (1954), I differ from them in assuming that feedback mechanisms are commoner than they think likely.

These are fairly basic differences in point of view; but rather than add to discussions such as those of Milne (1962), Bakker (1964), and Lack (1966), I shall review some of the methods that can be employed to decide empirically between rival points of view.

## SOME PROBLEMS OF TESTING

### Falsifiability

There are many reasons why it is difficult to decide between rival hypotheses. One of them is that if an event is considered in isolation, and explained without implications to other events, there is no way of disputing whatever interpretation the author cares to adopt (Medawar, 1957). Few ecologists state what they will accept as evidence against their point of view, some merely claim that a factor is "important" without describing the empirical procedures by which this belief may be tested. It is equally difficult to see how to test the claim of an author such as Huffaker (1966), who writes that "food may be a limiting and regulating factor for a given population even if utilization of supply at equilibrium is low." Few populations can be excluded from this category. There are also practical difficulties in testing theories that are satisfied by some percentage effect that is much smaller than the effect of the supposedly irrelevant variables.

If our aim is to produce some sort of unity out of the chaos of our field observations we must obviously get rid of superfluous hypotheses and stick to those that explain the greatest variety of instances. It is systematic simplicity of this sort that we should be searching for, rather than simplicity in the sense of familiarity. Pitelka (1958) misunderstands the principle of parsimony when he writes: "it may be a strain on Occam's razor to suggest genetical hypotheses regarding such fluctuations as long as more directly ecological explanations can be invoked and tested." Moreover these simpler explanations were invoked, tested, and rejected at an early stage of the enquiry into the fluctuations in numbers of small mammals (Summerhayes, 1941; Elton, 1942, 1955; Chitty, 1952, 1960).

Cole (1954) has another idea about simplicity: "The hypotheses of random population fluctuations has the further advantage of being probably the simplest possible explanation for population cycles." Simplicity in this sense is obtained by ignoring the associated biological changes and accepting any one of an infinite number of explanations for the purely numerical changes. Cole's interpretation may be justified, however, where population densities depart only slightly and irregularly from their mean (see below).

"When in doubt, appeal to experiment" wrote Eddington, and there can be few subjects to which this advice is more applicable than to population ecology. Varley (1957) and Varley and Gradwell (1963) discuss some of the possibilities and limitations of an experimental approach. We should be clear, however, that an experiment is a test of hypotheses and does not necessarily entail setting up artificial conditions (Medawar, 1957). The essence of an experiment is an active search for evidence against one's hypothesis; and though some implications are best observed in the laboratory, the logical form of a null hypothesis is unaffected by being applied to "mere observations" made under natural conditions (Lloyd, 1960). As a rule an investigator will wish to test a wide variety of predictions under both natural and artificial conditions, for no single type of result can be decisive, even when it goes against the hypothesis (Cohen and Nagel, 1934; Popper, 1959). The reason for this is that in any subject, but particularly in one as complex as population ecology, auxiliary hypotheses are also involved in the testing; thus authors wishing to protect their own point of view can always blame these other conditions for an otherwise contradictory result. The logical chain between hypothesis and observations should therefore be as simple as possible. Predictive equations (Morris, 1963) are not much use if they do not predict; fitting them retrospectively to the data merely tests the ingenuity of the author, not the empirical content of the model. "Irrefutability is not a virtue of a hypothesis (as some people seem to think), but a vice" (Popper, 1963).

## Irrelevant Variables

At an early stage in any investigation one's first task will usually be the purely empirical one of eliminating irrelevant variables. Having noticed an association between events we must find out whether or not this association is invariant. At least four possibilities must be recognized: (1) the association was mere co-incidence; (2) the events were concomitant effects of antecedent circumstances; (3) the antecedents were both necessary and sufficient for the phenomenon, or were (4) either necessary or sufficient but not both.

The first line of defense against faulty interpretations is to replicate the observations, varying the circumstances as much as possible. An investigation, if confined in its early stages to a detailed study of a single population, may be less instructive than a more superficial study of several. For preference both approaches should be used. Quite simple observations (e.g., Elton, 1955) often show that possibly relevant factors are not invariably associated with the phenomenon, a finding that saves one from building theoretical structures on unsound empirical foundations.

More troublesome than mere coincidence — for example, that of a hot summer with a rise (or fall) in numbers — are those possibly relevant factors that tend to vary, or necessarily vary, with a sharp rise or fall in numbers. Self-regulatory herbivore populations, as they become more abundant, will almost always have a smaller surplus of food; and as they become less abundant will usually provide a smaller surplus for their predators. More damage to the vegetation and a higher proportion of prey eaten by predators are thus among the expected consequences of this type of fluctuation. From correlations alone one could not decide between self-regulation and the rival interpretations that her-bivores increase until they starve (Lack, 1954), or alternatively that they are eaten before they do so (Hairston, Smith, and Slobodkin, 1960). A reduced food supply and a higher ratio of predators to prey will affect the course of events; but the inference that they, or a higher incidence of disease (Elton, 1942; Chitty, 1954), play any necessary part in them cannot be drawn from such data. Practical difficulties should not blind us to the limited explanatory value of studies which are not designed to distinguish between causes and effects.

Let us suppose, however, that we have found many instances in which some factor, process, or supposed cause (C) is present when the phenomenon, event, or supposed effect (E) is present, and which is absent when the phenomenon is absent. We then have the entries CE and $\overline{C}$E in two of the four possible cells of the following 2 × 2 contingency table (Chitty, 1954):

|  | | Supposed Cause (C) | |
|---|---|---|---|
|  | | Present | Absent |
| Supposed Effect (E) | Present | CE | $\overline{C}$E |
| | Absent | C$\overline{E}$ | $\overline{CE}$ |

To show that we have correctly identified a cause as both necessary and sufficient we must next produce evidence of failure to find it without its effect (C$\overline{E}$), or the effect without its supposed cause ($\overline{C}$E). If further study shows that C is not necessary for E we may be able to eliminate it from the relevant variables, or we may conclude that it is one of several that are sufficient to produce the effect (CE) without being necessary ($\overline{C}$E). This brings us up against the problem of the "plurality of causes."

To take an example: a change in food supply, predators, competitors, disease, weather, or any one of a long list of physical factors may be sufficient to alter the average level of abundance of a population; but none of them may be necessary. Again, let us suppose that a population declines and that the bare numerical change is the only observation recorded. Clearly anything may have caused the decline, and we can no more carry out a crucial test than we can diagnose a disease merely from knowing that a man's temperature has gone up. Yet doctors do diagnose a patient's troubles; they do so by taking account of his symptoms; and in general the solution to this well-known problem is through getting an accurate definition of the object or effect. A useful discussion is given by von Wright (1957). Suppose we wanted to check certain statements about phosphorus. Then, says von Wright: "The fundamental condition is that we have reliable criteria which enable us to decide when it is with a piece of phosphorus that we are dealing and when not." Cohen and Nagel (1934) put the matter as follows: "When a plurality of causes is asserted for an effect, the *effect* is not analyzed very carefully.

Instances which have significant differences are taken to illustrate the *same effect*. These differences escape the untrained eye, although they are noticed by the expert."

We can now see the peculiar merit of the typical decline in numbers of small mammal populations. This is a reasonably specific effect for which we are justified in trying to find a reasonably specific explanation; we can therefore eliminate hypotheses that fail to provide it. In a later section we shall see that, where populations are normally stationary, we do not have the same reliable criteria to help us when we are trying to distinguish between equally plausible but mutually incompatible hypotheses.

## Possibly Relevant Variables

Methods of elimination, while serving to narrow one's choice of alternatives, cannot be guaranteed to work as methods of discovery; and in claiming otherwise Platt (1964) bids us revive our faith in Baconian induction. While failures may spur invention, many of the best ideas owe little to logic; but by common consent we adopt a "didactic dead-pan" style (Watkins, 1964) and conceal the irrational ways in which we find the relevant variables and start getting positive results. The part played by luck, by errors that turned out to be useful, and by the work of colleagues and students, is an aspect of this story I can only briefly acknowledge.

A great deal of the work prior to 1959 was of the type illustrated in Figures 4 and 5, and is unlikely to be published; it consisted of unsuccessful attempts to find an association between population trends and changes in some organ or function that would indicate a pathological condition.

Figure 4 shows that two populations were out of phase from 1953 until 1957; Figure 5 shows that there was no difference between them in standardized mean adrenal weight — either in 1954, when numbers were low on one area and high on the other, or in 1955 when numbers were similar on both areas but going in opposite directions. Between years there was a difference in adrenal weight, which could have been misleading if either area

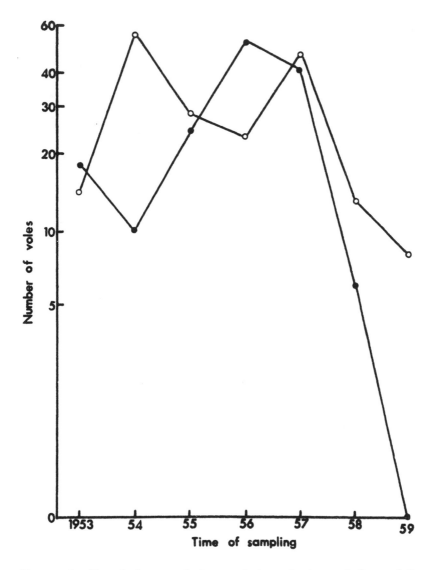

FIGURE 4: Population trends in two independent populations of the vole, *Microtus agrestis*, at Lake Vyrnwy, Wales, in April-May, 1953–59. The ordinate shows the number of voles taken per 100 traps in two days. (From Chitty and Chitty, 1962a. Scale: log n + 1.) o = Area F, • = Area R.

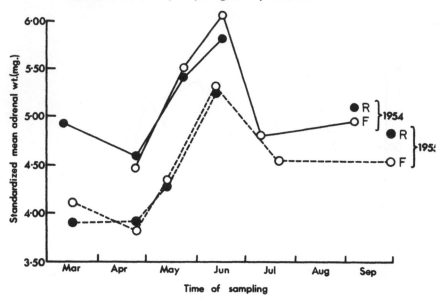

FIGURE 5: Seasonal and annual differences in standardized mean adrenal weights of male voles at Lake Vyrnwy, Wales, illustrating the need for comparative data to eliminate errors due to differences between years. Since the populations were out of phase (Figure 4) the null hypothesis was accepted that there was no relation between population trend and adrenal weight. (From H. Chitty, 1961.)

had been studied alone; seasonally there were differences that were probably due to changes in the amount of intraspecific strife (H. Chitty, 1961).

Although much time was spent in testing *mechanisms* that were wholly or partly false, there is still no reason to abandon the original *concept* that populations can regulate their own numbers through qualitative changes. The role of behavior in bringing about these changes was first studied in the Field Vole by Clarke (1953, 1956) and in the Partridge, *Perdix perdix*, by Jenkins (1961a, b); but the idea that behavior might be polymorphic is still too recent for us to know how fruitful it will be (Andrewartha, 1959; Krebs, 1964b; Chitty, 1964, 1965). Some of its implications have been looked into, however, and I shall refer briefly to a few of the results.

The most comprehensive studies have been carried out by

Krebs (1964b, 1966) on two species of lemmings in the Arctic and on *M. californicus*. He concluded that the observations to be explained were essentially the same as those previously described for *Microtus* spp. and were consistent with the scheme outlined in Figure 3. Skull proportions of the lemmings varied with the stage of their cycle in numbers, which also supported the idea that fundamental qualitative changes were taking place (Krebs, 1964a). Krebs himself did not study the behavior of his animals, though he noted that wounding occurred even at very low densities in lemmings, and refers to a study carried out on aggressive behavior in male *M. californicus*. These latter results did not turn out the way they were expected to (Krebs, pers. comm.), a warning that if behavior has the relevance predicted, we do not yet know what form it takes. Certainly we should not confine our observations to the grosser forms of aggression. Among other aspects the part played by odors in the defense of territories is likely to repay further study (Lyne, Molyneux, Mykytowycz, and Parakkal, 1964).

Other types of field experiments were carried out by Krebs (1966). An introduced population made rapid growth in an area on which a decline had occurred, which showed that the area itself was favorable. The introduced animals had been removed from an expanding population on another area, which confirmed one part of the prediction that "numbers should continue to increase if animals from an increasing population are successfully transferred to an area from which a declining population has been removed; but numbers should continue to decline if animals from a declining population are transferred to a new area" (Chitty, 1960).

However, when this prediction was made I thought that declining populations consisted predominantly of animals in which some adverse physiological change had occurred. Now that the change seems to consist of an increase in the proportion of aggressive animals it is less likely that a declining population would continue to decline if moved to a new area. Established social patterns might be sufficiently disrupted for results to go either way. Petruscewicz (1963) found that putting a stationary population of white mice into a different cage of the same size was sometimes enough to make their numbers go up.

Krebs (1966) also tested the idea that a heavily cropped population would retain the chief characteristics of an expanding population. Unfortunately, he could not prevent the numbers from being made up again by immigrants; but this at least showed that animals had previously avoided or been kept out of the area, and that the number of replacements was fairly predictable. Smyth (1966), in a similar attempt at cropping the Bank Vole (*Clethrionomys glareolus*) also found that in most months the animals he killed were almost all replaced by immigrants.

Sadleir (1965), using the deermouse, *Peromyscus maniculatus*, quantified the aggressive behavior of adult males and concluded that survival and recruitment of juveniles could be explained by seasonal changes in aggressive behavior of the adults. Both he and Britton (1966), by removing adults, increased the proportion of subadults reaching maturity. [Snyder (1961, 1962) had produced the same effect by cropping a population of woodchucks, *Marmota monax*.]

Healey (1966) tested Sadleir's conclusions by first establishing adult populations of aggressive and docile *Peromyscus* in the wild and then introducing a cohort of juveniles. There was usually a much greater loss of juveniles from the plots containing the aggressive adults, but Healey was unable to say that the aggressive and docile adults were genetically different.

The most crucial test of all has yet to be carried out (Chitty and Phipps, 1966, Appendix 1). According to hypothesis the animals present in stationary or declining populations have been selected for their superior ability to survive the effects of mutual interference. Animals in expanding populations, by contrast, are assumed not to have been so selected, but to be better fitted to withstand all other hazards of their environment. Therefore if animals of these different origins are placed together in the wild we should find the following differences in their respective contributions to the gene pool of later generations.

1. Let us first assume that the "environment of the population" remains favorable; then (a) at high population densities the

selective advantage should be with the supposedly aggressive individuals from the stationary or declining populations, but (b) the reverse should be true at very low population densities.

2. Now let us assume that there is a climatic catastrophe of the sort to which animals from the expanding population are supposed to be the more resistant. (a) According to the model previously discussed (Leslie, 1959), the direct effects of weather will be to remove a relatively low proportion of these resistant animals. However, the failure to find any evidence of reduced viability in voles now makes it less likely that weather has direct effects of the magnitude required. A second mechanism is therefore worth considering. (b) Let us suppose that the effect of the weather is on the habitat, which is made either more or less favorable. In relation to the new conditions, individuals will now find themselves either less crowded or more crowded, and may be expected to change their dispersion accordingly. This in turn may result in numbers either increasing or decreasing, or at least changing at a different rate from that expected.

Two vole populations that came into phase during the mild winter of 1956–57 may have followed this latter pattern; the denser population failed to decline until a year later than expected, while the expanding population reached its peak in three instead of the usual four years (Figure 4 and Chitty and Chitty, 1962a). Another mild winter, that of 1960–61, was also associated with an unexpected peak year instead of a further decline (Chitty and Phipps, 1966). These habitats were perhaps more than usually favorable when the surplus animals would normally have been eliminated.

The success of Krebs and Healey in setting up artificial populations in the wild suggests that we may now look forward to having populations out of phase when we want them. If so we can make an experimental attack on the problem of how weather affects the individuals, and what part it plays in the tendency of populations to keep in step. The idea that the effects of weather are independent of population density is so firmly rooted in ecological thinking that no one seems to have made the necessary observations to find out (cf. Moran, 1954).

The technique of setting up populations in the wild is not new: Einarsen (1945), Sheppard (1953, 1956), Sheppard and Cook (1962), and Miller (1958) are among those who have applied it; Morris (1960) quotes other examples.

## THE PROBLEM IN STATIONARY POPULATIONS

A frequent source of confusion is that the problem of accounting for differences in population density *between* areas is often identified with the problem of accounting for the relatively stationary state of populations *within* areas. One often meets the non sequitur that because parasites can reduce average population density, as in successful biological control, therefore they are also necessary to prevent such populations from reaching abnormally high densities. The problem of explaining differences in population density between areas, and of explaining the regulation of numbers within areas, may or may not be entirely separate problems; they certainly cannot be assumed to be the same, and the difficulty is to know how to study the variables relevant to one of these problems in spite of the uncontrolled variables which are relevant only to the other. The technical problem is much the same as that encountered in agricultural field trials of removing effects due to differences between areas from effects within areas due to treatments or varieties of crop. "It would be common for the yields on individual plots in a field to vary by as much as ± 30 per cent from their mean, and a systematic difference of 5 per cent between varieties might be of considerable practical importance. We shall be concerned with methods for arranging the experiment so that we may with confidence and accuracy separate the varietal differences, which interest us, from the uncontrolled variations, which do not" (Cox, 1958).

In population ecology we are of course interested in differences in abundance both between and within areas. We need to know why, on the average, population densities are higher in some areas than in others, i.e., why one environment is more or less favorable than another, or what change is sufficient to make it so

(Morris, 1963; Geier, 1966). In either case the answer can be sought in comparative terms: by finding out what factors are associated with observed differences in population density, or what changes occur when a factor is artificially varied. In the present context, however, this problem concerns us only to the extent already discussed: that uncontrolled changes in certain aspects of the environment make conditions either more or less favorable. Some population changes are adjustments to such environmental changes; others are not; the difficulty is to tell one from the other. If we cannot distinguish between different sorts of numerical change we cannot draw valid inferences about regulatory mechanisms; on the other hand if populations do not fluctuate at all we have no worthwhile comparative data whatever.

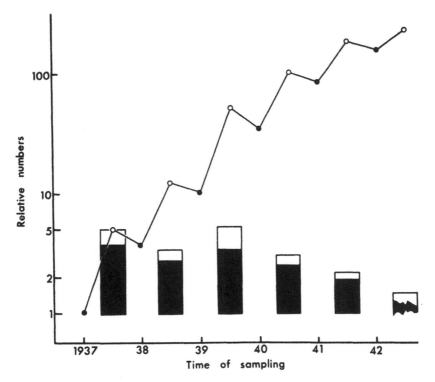

FIGURE 6: · Estimated number of Pheasants (*Phasianus colchicus*) on Protection Island, Washington, in March and November 1937–42 ( • ). Histograms show the relative size of spring and fall populations, and (in black) the net annual rates of increase. (From Einarsen, 1945.)

Figure 6 shows one of the relatively few instances in which a pronounced increase in numbers has been measured on a population released under natural conditions. Eight artificially reared Pheasants, *Phasianus colchicus*, liberated on an island in the state of Washington, increased within six seasons to the abnormally high density of about nine birds per suitable acre (Einarsen, 1945). Had the population become stationary, as it seemed about to, the problem would have been to compare this regulated state with the previous one, which must have been as near to being unregulated as one is likely to find. And had the population subsequently declined of its own accord we might have found an even more obvious contrast with its unregulated state. As it happened, however, the environment was rudely disturbed, and the decline that occurred was irrelevant to the problem of natural regulation.

The histograms show the net annual increase expressed as the

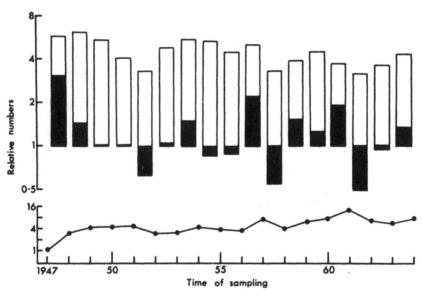

FIGURE 7: Estimated number of Great Tits (*Parus major*) breeding in Marley Wood Oxfordshire in 1947–65 ( ● ). Histograms show the relative sizes of spring and summer populations, and (in black) the net annual changes in numbers. The breeding population varied between 14 in 1947 and 172 in 1961. (From Lack, 1966, pp. 60–61.)

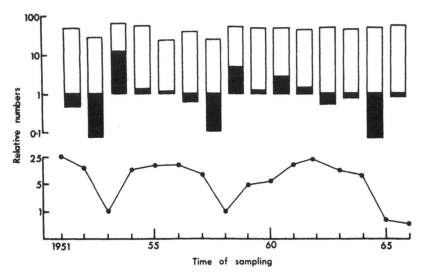

FIGURE 8: Estimated numbers of adult Pine Loopers (*Bupalus pini-arius*) in De Hoge Veluwe, the Netherlands in June 1951–66 (•). Histograms show the relative sizes of adult and egg populations, and (in black) the net annual changes in numbers. The adult population varied between 2.7 per sq. m. in June 1950 and 0.05 (max.) per sq. m. in June 1966. Unity = 0.1 per sq. m. (From Klomp, 1966, pp. 268–70, and personal communication for data for 1965–66.)

difference between the proportionate increase and decrease. Between spring and fall these birds were capable of a fivefold increase; between fall and spring their decrease was always slight. The reduced rates of increase after 1939 were thus due to factors that interfered with recruitment during the summer.

Figures 7 and 8 offer a sharp contrast to the previous example in showing how seldom these particular populations made any appreciable departure from a relatively stationary state. The breeding population of Great Tits (Lack, 1966) increased three-fold between 1947 and 1948 and doubled between 1956 and 1957; all other changes were between + 69 per cent and − 50 per cent of the numbers breeding the previous year. The adult population of Pine Loopers (*Bupalus piniarius*) also remained fairly stationary (Klomp, 1966): by comparison with an elevenfold increase in 1953–54 and an almost fivefold increase in 1958–59, all other increases were slight. These studies thus provide many examples of the phenomenon — a more or less stationary state — but few comparative observations on the state of rapid expansion from low numbers. It is not surprising, therefore, to find Lack (1966, p. 26) dismissing the high clutch sizes of 1947 and 1948 as abnormal, or including them with data collected at later stages of population growth, which make the averages "more reliable."

In both studies reproduction was adequate throughout, and the problem is to explain why the progeny failed to survive in sufficient numbers to produce a continual expansion, comparable to that observed among Einarsen's pheasants. Lack believed the Great Tits lost their young because of starvation, but was unable to find any evidence. Klomp believed that larval interference in *Bupalus* was responsible for reducing viability in the next generation; but even before getting the unpublished evidence for 1965–66 (Figure 8) he was unable to find a statistically significant relation between larval density in one year and egg or juvenile mortality the next.

After the mild winters of 1956–57 and 1960–61, not only in England but also in Holland (Lack, 1966), breeding populations of Great Tits were higher than expected; certain *Microtus* populations were also unexpectedly high in the spring of these same years (p. 14). Unfortunately we do not know whether or not these associations of events were fortuitous.

According to Klomp (1966): "The study of the population dynamics of insects has progressed further than that of any other animal group excepting perhaps the birds." Klomp does not suggest how this belief should be tested; but an appropriate null hypothesis might be that there is no difference between ecologists in their ignorance of why populations behave as they do.

I will conclude this section by referring to another study carried out on a normally stationary class of populations. Smith (1965) carried out a two-year study on chickadees (*Parus atricapillus*), a resident species usually found in fair numbers on the southwest coast of British Columbia. In this study the habitat was chosen in advance as likely to be favorable for their survival and reproduction; it was thus expected that the effects to be measured would be large in comparison with those of the irrelevant variables. The area was residential, partly surrounded by woods; it provided good cover, nesting sites, and food — the latter supplemented by several feeding tables maintained throughout the winter. Little predation seemed likely; only one kill was reported. The birds were nearly all colorbanded, and counts were made of the numbers alive at two-weekly intervals.

During fall and winter an average of over 95 per cent of the

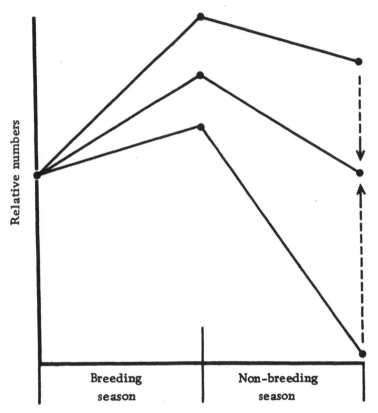

FIGURE 9: Seasonal increase and decrease of stationary populations living under hypothetically favorable, unfavorable, and intermediate conditions. Self-regulatory behavior might be obvious only if good reproduction and survival were followed by a sudden drop in numbers (cf. Jenkins, 1961a, Smith, 1965), though not necessarily at the time shown. Under other conditions numbers might be less obviously regulated through behavior restricting the amount of immigration.

birds present at the start of each two-week period were alive at the start of the next; but during a single period in February 1964 and again in March 1965, when the flocks broke up into pairs, only 60–70 per cent of the birds remained on the area. No such sharp change in numbers could have been expected, however, if reproduction had been less successful or if the autumn and winter survival rate had been just slightly lower (90 per cent period without immigration); and in less favorable habitats there might even have been an increase in numbers after the break-up of flocks on better areas (Figure 9).

These observations are an example of the self-regulatory behavior to be expected in a habitat where the chances of surviving the purely contingent mortality factors are extremely good. But because individuals of this species are capable of spacing themselves out in a way that limits their breeding density one does not infer that all populations of chickadees are self-regulating. Those that are and those that are not must be determined empirically (cf. Kluyver and Tinbergen, 1953); predictions appropriate to the testing of various other hypotheses must then be stated in their own operational terms. These other classes of population possibly do behave in the manner postulated by authors such as Nicholson (1954), Hairston, Smith, and Slobodkin (1960), or Lack (1966); but the relation of a population to its requisites cannot be deduced from unsubstantiated principles (Murdoch, 1966).

## DISCUSSION

The principal claim in this paper is that population densities, while obviously varying in response to numerous environmental factors, are further affected by a continual process of selection. If we can demonstrate that interference has severe effects on survival and reproduction it is less reasonable to assume that all genotypes suffer equally than to assume they are affected non-randomly. Selective advantage is measured in terms of *relative* contribution to the gene pool of later generations; hence a form of behavior that reduces an individual's absolute contribution would nevertheless be selected for if at the same time it reduced still further the contribution of other individuals. This explanation certainly does less violence to present ideas about natural selection than does Wynne-Edwards' view (1962) that reduced fecundity has been evolved because of its advantages to the population. Dispersion, then, and its effect on population density, should be explained in the ultimate terms of its evolutionary significance, not in terms of its proximate advantages, either to the individual or the group.

These proximate advantages are sometimes obvious — less crowding means more food, for example — but we cannot infer that dispersion mechanisms have been selected for this sort of reason, a point emphasized by Hinde (1956). Territorial behavior is the most obvious of these mechanisms, but Lack (1954, p. 263) doubts that territorial behavior "is important in maintaining a food supply or in limiting numbers." "If territorial behavior as such limits population density" he writes on page 260, "then the size of the territories should be nearly constant in each species." Unfortunately he does not explain why size of territory should differ from most other expressions of activity in being unaffected by environmental variables.

Territorial behavior probably has many functions that are not selected for, and in some instances perhaps its only selective advantage is in excluding other animals from the best breeding places. Brown (1964) seems to hold a minority view when he writes: "As long as counter selection against aggressiveness were weak, *aggressiveness per se would be maintained in the population merely by the exclusion of less aggressive birds from breeding.*"

In some species the selective advantages of dispersion may be confined to a short period in the life of the organism, which for the rest of the time may lead a social life (Barnett, 1964; Brereton, 1966) and be subjected to strong selective pressures of other sorts. The effects of selection should also vary according to the habitat, owing to differences in the proportions eliminated by intraspecific behavior (Chitty, 1965, Figure 1). These expected differences might be reduced by gene flow, although immigrants to the worst habitats should consist mainly of animals selected for avoidance of the more crowded areas (Lidicker, 1962). During a general decline in their numbers, Bank Voles (*Clethrionomys glareolus*) survived better in the habitats that had been least crowded (Evans, 1942). Such habitats, Evans suggested, might be essential for the survival of the species.

The sudden removal of a mortality factor, such as an effective parasite, that had minimized the frequency of intraspecific interactions, should be followed by unusually good survival, high abundance, and damage to resources. Subsequent peaks should be appreciably lower after the strong selection presumably exerted during the first decline. Similar results might occur after the introduction of animals raised in captivity, where wildness is presumably selected against. Outbreaks might also occur after any long period of expansion into a relatively empty universe. Such opportunities often follow invasions; they should also follow severe catastrophes other than those due to the animal's own behavior. Watt (1963) recognizes that genetic changes may have occurred after DDT spraying against Spruce Budworm (*Choristoneura fumiferana*): "A striking feature of budworm survival data has been the rise in survival the season after a time of great adversity for budworm. It does not seem to matter what causes the adversity.... Two examples of post-adversity rises in survival ... one resulting from spraying and the other from starvation, are typical of many in our data." The unusual abundance of pheasants observed by Einarsen (1945) may have exemplified the combined effects of previous domestication and a long period of expansion in the absence of wild residents (cf. Miller, 1958).

Some years ago Ford (1955) wrote that ecologists all too often "deal with what are essentially genetic phenomena but rarely attempt to investigate them genetically." There is much justice in this criticism, though it is equally clear that the traffic in ideas must flow both ways. Birch (1960) and Milne (1962) summarize many of the ideas of ecologists about the role of selection in determining animal numbers; two more recent papers are those of Wilbert (1963) and Pimentel (1964).

Studies on the pathological effects of crowding, reviewed or carried out by Christian and Davis (1964), and Myers (1966), show that the "environment of the individual" can be very harsh in artificially crowded populations. Many of these effects are probably gross exaggerations of one part of what goes on in nature, or are complete artifacts. Nevertheless, given field evidence of unexplained reductions in breeding success and survival, we can be reasonably sure that these experimental results indicate that powerful intraspecific processes of some kind are at work in the wild. If so, the chances of them being non-selective seem remote, and we may expect population parameters to vary according to the number of generations exposed to these stresses.

It seems, then, that behavior, physiology, and genetics must be of increasing concern to population ecologists, who have probably spent too long already on purely descriptive studies.

## SUMMARY

1. Mechanisms for the self-regulation of animal numbers are thought to be a consequence of selection, under conditions of mutual interference, in favor of genotypes that have a worse effect on their neighbors than vice versa. This idea is highly speculative and is therefore presented in conjunction with operational tests of its implications. Evidence for the idea is reviewed only briefly, the aim of the paper being to elaborate the consequences of this hypothesis to make it more readily falsifiable.

2. Populations of small mammals, during their periodic declines in numbers, present a clearly recognizable phenomenon, whose most puzzling feature is the severity of the losses at relatively low densities after peak abundance. There is no evidence that the animals are less viable than normal, though a decrease in some components of fitness may reasonably be inferred.

3. For a crucial test of these ideas it will be necessary to measure the selective advantage of the supposed behavioral polymorphs — when they occur together both in crowded

and uncrowded conditions and exposed both to good and to bad weather. Various workers have now quantified behavior and made preliminary tests on populations specially established in the wild.

4. In populations that remain fairly stationary, many slight changes in numbers are due to changes in variables that are irrelevant to the problem of what prevents unlimited increase — an example of the need for comparative observations between any phenomenon and its control. Instances of rapid expansion in numbers sometimes provide the requisite contrasts with the stationary state.

5. Since we cannot yet analyze all stationary populations into recognizable classes we do not have specific phenomena whose explanation we can test by prediction, nor can we unify our explanations for differences in abundance between areas. However, in habitats that temporarily become more or less favorable than usual (e.g., through changes in weather) there may be associated changes in the amount of crowding that animals will tolerate. Such changes might explain why some populations tend to fluctuate in step.

6. Individuals that have not been selected under conditions of intraspecific strife are likely to be mutually tolerant, and populations composed of such individuals should reach abnormal abundance before such selection takes place. If conditions are otherwise favorable we might find such populations (a) when new areas are invaded, especially if (b) the animals released into a vacant niche are from stocks reared under conditions that select against aggressive behavior; (c) when an existing source of heavy mortality is removed; (d) when populations have been seriously reduced by natural or artificial catastrophes.

7. These ideas, whether themselves true or false, may suggest further work on behavior, physiology, and genetics that will contribute to the solution of ecological problems.

# REFERENCES

Andrewartha, H. G. (1959). Self-regulatory mechanisms in animal populations. *Aust. J. Sci.*, 22:200–205.

——, and Birch, L. C. (1954). "The distribution and abundance of animals." (Univ. Chicago Press.)

Bakker, K. (1964). Backgrounds of controversies about population theories and their terminologies. *Z. ang. Ent.*, 53:187–208.

Barnett, S. A. (1964). Social stress. *Viewpoints in Biol.*, 3:170–218.

Birch, L. C. (1960). The genetic factor in population ecology. *Amer. Nat.*, 94:5–24.

Brereton, J. Le Gay (1962). Evolved regulatory mechanisms of population control. In "The evolution of living organism," ed. G. W. Leeper (Melbourne Univ. Press), pp. 81–93.

——. (1966). The evolution and adaptive significance of social behaviour. *Proc. ecol. Soc. Aust.*, 1:14–30.

Britton, M. M. (1966). Reproductive success and survival of the young in *Peromyscus*. M.Sc. Thesis, Univ. British Columbia.

Brown, J. L. (1964). The evolution of diversity in avian territorial systems. *Wilson Bull.*, 76:160–169.

Chitty, D. (1952). Mortality among voles (*Microtus agrestis*) at Lake Vyrnwy, Montgomeryshire in 1936–39. *Phil. Trans. B.*, 236:505–552.

——. (1954). Tuberculosis among wild voles: with a discussion of other pathological conditions among certain mammals and birds. *Ecology*, 35:227–237.

——. (1955). Adverse effects of population density upon the viability of later generations. In "The numbers of man and animals," eds. J. B. Cragg and N. W. Pirie. (Oliver & Boyd, Edinburgh), pp. 57–67.

——. (1958). Self-regulation of numbers through changes in viability. *Cold Spring Harbor Symp. Quant. Biol.*, 22:277–280.

——. (1960). Population processes in the vole and their relevance to general theory. *Canad. J. Zool.*, 38:99–113.

——. (1964). Animal numbers and behaviour. In "Fish and wildlife: A memorial to W. J. K. Harkness," ed. J. R. Dymond. (Longmans, Canada), pp. 41–53.

——. (1965). Predicting qualitative changes in insect populations. *Proc. 12th Int. Congr. Ent. Lond.*, 384–386.

——, and Chitty, H. (1962a). Population trends among the voles at Lake Vyrnwy, 1932–60. *Symp. Theriologicum, Brno. 1960*:67–76.

——, and Phipps, E. (1966). Seasonal changes in survival in mixed populations of two species of vole. *J. Anim. Ecol.*, 35:313–331.

Chitty, H. (1961). Variations in the weight of the adrenal glands of the field vole, *Microtus agrestis*. *J. Endocrin.*, 22:387–393.

——, and Chitty, D. (1962b). Body weight in relation to population phase in *Microtus agrestis*. *Symp. Theriologicum, Brno. 1960*:77–86.

Christian, J. J., and Davis, D. E. (1964). Endocrines, behaviour, and population. *Science*, 146:1550–1560.

——, Lloyd, J. A., and Davis, D. E. (1965). The role of endocrines in the self-regulation of mammalian populations. *Recent. Progr. Hormone Res.*, 21:501–571.

Clarke, J. R. (1953). The effect of fighting on the adrenals, thymus and spleen of the vole (*Microtus agrestis*). *J. Endocrin.*, 9:114–126.

——. (1956). The aggressive behaviour of the vole. *Behaviour*, 9:1–23.

Cohen, M. R., and Nagel, E. (1934). "An introduction to logic and scientific method." (Routledge, London.)

Cole, L. C. (1954). Some features of random population cycles. *J. Wildlife Manag.*, 18:2–24.

Cox, D. R. (1958). "Planning of experiments." (Chapman & Hall, London.)

Davis, D. E. (1951). The relation between level of population and pregnancy of Norway rats. *Ecology*, 32:459–461.

Einarsen, A. S. (1945). Some factors affecting ring-necked pheasant population density. *Murrelet*, 26:2–10; 39–44.

Elton, C. S. (1931). The study of epidemic diseases among wild animals. *J. Hyg., Camb.*, 31:435–456.

———. (1942). "Voles, mice and lemmings: problems in population dynamics." (Clarendon Press, Oxford.)

———. (1955). Discussion. In "The numbers of man and animals," eds. J. B. Cragg & N. W. Pirie. (Oliver & Boyd, Edinburgh), pp. 82–83.

Evans, F. C. (1942). Studies of a small mammal population in Bagley Wood, Berkshire. *J. Anim. Ecol.*, 11:182–197.

Ford, E. B. (1955). "Moths." (Collins, London.)

———. (1961). "Mendelism and evolution." 7th ed. (Methuen, London.)

———. (1964). "Ecological genetics." (Methuen, London.)

Geier, P. W. (1966). Management of insect pests. *Ann. Rev. Ent.*, 11:471–490.

Godfrey, G. K. (1955). Observations on the nature of the decline in numbers of two *Microtus* populations. *J. Mammal.*, 36:209–214.

Green, R. G., and Evans, C. A. (1940). Studies on a population cycle of snowshoe hares on the Lake Alexander Area. *J. Wildlife Manag.*, 4:220–238, 267–278, 347–358.

Hairston, N. G., Smith, F. E., and Slobodkin, L. B. (1960). Community structure, population control, and competition. *Amer. Nat.*, 94:421–425.

Hamilton, W. J. (1937). The biology of microtine cycles. *J. agric. Res.*, 54:779–790.

———. (1941). Reproduction of the field mouse *Microtus pennsylvanicus* (Ord). Mem. Cornell Univ. agric. exp. Sta. No. 237.

Healey, M. C. (1966). Aggression and self-regulation of population size in deermice. M.Sc. thesis, Univ. British Columbia.

Hinde, R. A. (1956). The biological significance of the territories of birds. *Ibis*, 98:340–369.

Huffaker, C. B. (1966). Competition for food by a phytophagous mite: the roles of dispersion and superimposed density-independent mortality. *Hilgardia*, 37:533–567.

Jenkins, D. (1961a). Population control in protected partridges (*Perdix perdix*). *J. Anim. Ecol.*, 30:235–258.

———. (1961b). Social behaviour in the partridge, *Perdix perdix*. *Ibis*, 103 A:155–188.

Kalela, O. (1957). Regulation of reproduction rate in subarctic populations of the vole *Clethrionomys rufocanus* (Sund.). *Ann Acad. Sci. Fenn. Ser. A., Sect. 4, No. 34.*

Klomp, H. (1964). Intraspecific competition and the regulation of insect numbers. *Ann. Rev. Ent.*, 9:17–40.

———. (1966). The dynamics of a field population of the pine looper, *Bupalus piniarius* L. (Lep., Geom.) *Adv. Ecol. Res.*, 3:207–305.

Kluyver, H. N., and Tinbergen, L. (1953). Territory and the regulation of density in titmice. *Archs. Néerl. Zool.*, 10:265–289.

Koskimies, J. (1955). Ultimate causes of cyclic fluctuations in numbers in animal populations. *Papers Game Res., Helsinki*, 15:1–29.

Krebs, C. J. (1964a). Cyclic variation in skull-body regressions of lemmings. *Canad. J. Zool.*, 42:631–643.

———. (1964b). The lemming cycle at Baker Lake, Northwest Territories, during 1959–62. Arctic Inst. N. Amer. Tech. Paper No. 15.

———. (1966). Demographic changes in fluctuating populations of *Microtus californicus*. *Ecol. Monog.*, 36:239–273.

Lack, D. (1954). "The natural regulation of animal numbers." (Clarendon Press, Oxford.)

———. (1966). "Population studies of birds." (Clarendon Press, Oxford.)

Leslie, P. H. (1959). The properties of a certain lag type of population growth and the influence of an external random factor on a number of such populations. *Physiol. Zool.*, 32:151–159.

Lidicker, W. Z. (1962). Emigration as a possible mechanism permitting the regulation of population density below carrying capacity. *Amer. Nat.*, 96:29–33.

Lloyd, M. (1960). Statistical analysis of Marchant's data on breeding success and clutch-size. *Ibis*, 102:600–611.

Lyne, A. G., Molyneux, G. S., Mykytowycz, R., and Parakkal, P. F. (1964). The development, structure and function of the submandibular cutaneous (chin) glands in the rabbit. *Aust. J. Zool.*, 12:340–348.

Mather, K. (1961). Competition and co-operation. *Symp. Soc. exp. Biol.*, 15:264–281.

Medawar, P. B. (1957). "The uniqueness of the individual." (Methuen, London.)

Miller, R. B. (1958). The role of competition in the mortality of hatchery trout. *J. Fish. Res. Bd. Canada*, 15:27–45.

Milne, A. (1962). On a theory of natural control of insect population. *J. Theoret. Biol.*, 3:19–50.

Moran, P. A. P. (1954). The logic of the mathematical theory of animal populations. *J. Wildlife Manag.*, 18:60–66.

Morris, R. F. (1960). Sampling insect populations. *Ann. Rev. Ent.*, 5:243–264.

———. (1963). Predictive population equations based on key factors. *Mem. Ent. Soc. Canada*, 32:16–21.

Murdoch, W. M. (1966). "Community structure, population control, and competition"—a critique. *Amer. Nat.*, 100:219–226.

Myers, K. (1966). The effects of density on sociality and health in mammals. *Proc. ecol. Soc. Aust.*, 1:40–64.

Nelson, J. B. (1964). Factors influencing clutch-size and chick growth in the North Atlantic gannet, *Sula bassana. Ibis*, 106:63–77.

———. (1965). The behaviour of the gannet. *Brit. Birds*, 58:233–288; 313–336.

Newson, J., and Chitty, D. (1962). Haemoglobin levels, growth and survival in two *Microtus* populations. *Ecology*, 43:733–738.

Newson, R. (1963). Differences in numbers, reproduction and survival between two neighbouring populations of bank voles (*Clethrionomys glareolus*). *Ecology*, 44:110–120.

Nicholson, A. J. (1954). An outline of the dynamics of animal populations. *Aust. J. Zool.*, 2:9–65.

Petruscewicz, K. (1963). Population growth induced by disturbance in the ecological structure of the population. *Ekol. Polska A*, 11:87–125.

Pimentel, D. (1964). Population ecology and the genetic feed-back mechanism. In "Genetics Today": Proc. 11th Int. Congr. Genetics, pp. 483–488.

Pitelka, F. A. (1958). Some aspects of population structure in the short-term cycle of the brown lemming in northern Alaska. *Cold Spr. Harb. Symp. Quant. Biol.*, 22:237–251.

Platt, J. R. (1964). Strong inference. *Science*, 146:347–353.

Popper, K. R. (1959). "The logic of scientific discovery." (Hutchinson, London.)

———. (1963). "Conjectures and refutations: the growth of scientific knowledge." (Routledge and Kegan Paul, London.)

Sadleir, R. M. F. S. (1965). The relationship between agnostic behaviour and population changes in the deermouse, *Peromyscus maniculatus* (Wagner). *J. Anim. Ecol.*, 14:331–352.

Sheppard, P. M. (1953). Evolution in bisexually reproducing organisms. In "Evolution as a process," eds. J. S. Huxley, A. C. Hardy, and E. B. Ford. (Allen & Unwin, London), pp. 201–218.

———. (1956). Ecology and its bearing on population genetics. *Proc. Roy. Soc. B.*, 145:308–315.

———, and Cook, L. M. (1962). The manifold effects of the *medionigra* gene in the moth *Panaxia dominula* and the maintenance of a polymorphism. *Heredity*, 17:415–426.

Smith, S. M. (1965). Seasonal changes in the survival of the black-capped chickadee. M.Sc. Thesis, Univ. British Columbia.

Smyth, M. (1966). Winter breeding in woodland mice, *Apodemus sylvaticus*, and voles, *Clethrionomys glareolus* and *Microtus agrestis*, near Oxford. *J. Anim. Ecol.*, 35:471–485.

Snyder, R. L. (1961). Evolution and integration of mechanisms that regulate population growth. *Proc. Nat. Acad. Sci.*, 47:449–455.

———. (1962). Reproductive performance of a population of woodchucks after a change in sex ratio. *Ecology*, 43:506–515.

Summerhayes, V. S. (1941). The effect of voles (*Microtus agrestis*) on vegetation. *J. Ecol.*, 29:14–48.

Tinbergen, N. (1957). The functions of territory. *Bird Study*, 4:14–27.

Varley, G. C. (1957). Ecology as an experimental science. *J. Anim. Ecol.*, 26:251–261.

———, and Gradwell, G. R. (1963). The interpretation of insect population changes. *Proc. Ceylon Ass. Adv. Sci.*, 1962, 18:142–156.

Watkins, J. W. W. (1964). Confession is good for ideas. In "Experiment: a series of scientific case histories . . . ," ed. D. Edge. (B.B.C., London), pp. 64–70.

Watt, K. E. F. (1963). The analysis of the survival of large larvae in the unsprayed area. *Mem. Ent. Soc. Canada*, 31:52–63.

Wellington, W. G. (1960). Qualitative changes in natural populations during changes in abundance. *Canad. J. Zool.*, 38:289–314.

Wilbert, H. (1963). Können Insektenpopulationen durch Selektionsprozesse reguliert werden? *Z. ang. Ent.*, 52:185–204.

von Wright, G. H. (1957). "The logical problem of induction." (Blackwell, Oxford.)

Wynne-Edwards, V. C. (1962). "Animal dispersion in relation to social behaviour." (Oliver & Boyd, Edinburgh.)

# 10. Population Regulation and Genetic Feedback

## DAVID PIMENTEL

Although within a relatively short period man has learned how to put himself into space, he still is not certain how the numbers of a single plant or animal population are naturally controlled. Aspects of this problem have been investigated since Aristotle's time, they were given important consideration in Darwin's *Origin of Species*, and yet the unknowns far outweigh the discoveries. If we knew more about natural regulation of population, we would be in a better position to devise more effective and safer means of control for important populations of plant and animal pests. We might also be better able to limit the growth of human populations, although that problem is exceedingly complex because of the social activities and nature of man.

From *Science*, 159 (March 29, 1968), 1432–1437. Copyright 1968 by the American Association for the Advancement of Science. With the author's permission, the first figure of the original paper, showing a wasp parasite and its housefly host pupa, has been omitted, and the remaining figures are renumbered accordingly. The author acknowledges support in part by environmental biology grant GB–4567 from NSF.

## POPULATION CHARACTERISTICS

Before considering how populations in nature are regulated, we should review various characteristics of animals and plants — as individuals and as populations. Do populations of animals in nature fluctuate severely or are they relatively constant? Stability and constancy have been proposed as characteristics of natural populations. Speaking about birds, Lack (1) says, "of the species which are familiar to us in England today, most were familiar to our Victorian great-grandparents and many to our medieval ancestors; and the known changes in numbers are largely attributable to man." He continues, "All the available censuses confirm the view that, where conditions are not disturbed, birds fluctuate in numbers between very restricted limits. Thus, among the populations considered above, the highest total recorded was usually between two and six times, rarely as much as ten times, the lowest. This is a negligible range compared with what a geometric rate of increase would allow." Discussing the stability in animal populations in general, MacFadyen (2) writes: "it is generally agreed that the same species are usually found in the same habitats at the same seasons for many years in succession, and that they occur in numbers which are of the same order of magnitude."

Further evidence for the thesis that species populations are relatively constant is found in a study of the changes in the fauna of Ontario, Canada (3). When Snyder (4) evaluated the bird fauna, he found that, over a period of about 70 years, two species became extinct, 23 species increased in number, and 6 species decreased in number. This represents a total change of only 9 per cent of 351 bird species found in Ontario (5) and agrees favorably with an 11 per cent change (6) for 149 species of birds over a 50-year period in Finland. These data suggest that there is relative constancy in the abundance of species populations. The word "relative" must be emphasized because changes in numbers must be related to a species' real potential for fluctuations; to paraphrase Lack (1), the changes observed are mere "ripples" compared to the possible "waves." Although in geological time 99 per cent of all species have become extinct, during periods of 100 years or more constancy is the rule.

There are exceptions to this rule of constancy. What are the population characteristics of plants and animals newly introduced on islands and continents? Typically when a species population enters a new biotic community in which no ecological barrier exists, outbreaks occur in these populations. The following examples of introductions into the United States illustrate this point: Japanese beetle (*Popilla japonica*); European gypsy moth (*Porthetria dispar*); South American fire ant (*Salenopsis saevissima*); Asiatic chestnut blight (*Endothia parasitica*) (fungus); European starling (*Sturnus vulgaris*); and the English sparrow (*Passer domesticus*). Outbreak of chestnut blight was so severe that for all practical purposes it destroyed its host. the American chestnut tree (*Castanea dentata*). After increases in the number of Japanese beetles, a bacterial pathogen epidemic spread through the population and is now effectively controlling the numbers of the beetle.

In nature the numbers of many herbivore. parasitic. and predaceous species are limited by resistant factors inherent in the host. Are resistant factors which limit or prevent pest attack commonly found in plants and animals? Various kinds of resistant factors exist in plants and animals in nature and appear to be quite prevalent. The spines occurring in many kinds of plants. such as cacti, gorse, and hawthorn, prevent feeding by browsing animals. Toxins or growth inhibitors which occur in many kinds of plants limit animal feeding, for example, tannins in oak leaves (7), cyanide in bird's-foot trefoil (*Lotus corniculatus*) (8). and nepetalactone in catnip (9). To prevent predator attack (10). poisonous sprays are ejected from many insects and other arthropods, such as acetic acid by whip scorpion (Pedipalpidas). formic acid by ants (Formicidae), and *p*-benzoquinones by flour beetles (Tenebrionidae). Repellent sprays from glands in some vertebrates, such as the skunk (*Mephitis* spp.), the Indian mongoose (*Herpestes auropunctatus*), and the toad (*Bufo marinus*), ward off attacking enemies. Nutritional changes in certain plants prevent the multiplication of attacking insects, for example, aphids on corn (carotene) (11) and leafhoppers on beets (linoleic acid) (12). A kind of armor plating protects various animals (armadillos, turtles, and certain beetles) from their attackers. Such physiological defense mechanisms as specific antibodies and phagocytosis are present in many kinds of animals (man and other vertebrates) and effectively control pathogen and parasite infections.

When these natural resistant factors in plants and animals are successful, they prevent the uncontrolled increase of the feeding species. Are animal numbers abundant or rare? Rarity, like constancy, is relative. Numbers of a given species can be related to the numbers of another species, to the unit area occupied, or to the food resources of the species. Andrewartha and Birch (13) noted that "the truth is that the vast majority of species are rare, by whatever criterion they are judged." In the *Origin of Species*, Darwin wrote, "rarity is the attribute of a vast number of species of all classes, in all countries." In enumerating the number of insects abundant enough to be considered pests, Smith (14) warned "that such species form only an insignificant fraction of the total number of phytophagous insects." Of the 240 species of nocturnal Lepidoptera collected by Williams (15), 35 species were represented by a single individual each; 85 (including the 35 above) were represented by 5 or fewer individuals; 115 by 10 or fewer; and 205 by 100 or fewer individuals; therefore, there were only 35 species with over 100 individuals. Further data in support of rarity are found in Dunn's (16) work with Panamanian snakes; he reports that "about $\frac{1}{10}$ of the species make up $\frac{1}{2}$ of the individuals in the snake populations."

Many of the abundant species would not be classed as abundant if they were compared to their food source. For example, many species of insects that are easily captured in the field are rare if they are sought on their host plant or if their biomass is compared with the biomass of the plant or animal upon which they feed.

One of the dynamic relationships in the community and ecosystem is the food chain, because animals must seek food to live. Elton (17) stated that the "whole structure and activities of the community are dependent upon questions of food-supply." What proportion of animals feed on living, as opposed to nonliving. matter? In nature, the majority of all animals may be classified as herbivore (grazer on living plants), parasite, or predator; few species are truly saprophytic. Though many animals are associated with dead plant matter, these animals are not saprophytes but are herbivores feeding on bacteria, fungi, and other minute organisms in the decaying matter. Jacot (18) stated, "I am quite certain that perhaps as much as one half of the Oribatoidea are not saprophytic. Their function is feeding on fungi." Overgaard (19) reported that the evidence suggested that nematodes feed not upon humus but upon plant roots, fungi, bacteria, and other animals. Speaking similarly about saprophagous insects. Chapman (20) noted that, "These are usually designated as those feeding upon decaying and fermenting matter. It is evident at the start that these insects live in media which may be teeming with microorganisms. and that the decaying material is the medium upon which the microorganisms live." *Drosophila* depend upon yeasts and other microorganisms present in decaying fruit (21).

## GENETIC FEEDBACK

Stability and constancy are characteristics of natural populations: in many hosts there are resistant factors that limit any severe attack of feeding species. and most animals feed on living matter. These seemingly diverse factors are related and are the foundation of the mechanism for population regulation which I termed "genetic feedback" (22). Population numbers (herbivore. parasite. or predator) are regulated in this way: high herbivore densities create strong selective pressures on their host-plant populations; selection alters the genetic makeup of the host population to make the host more resistant to attack; this in turn feeds back negatively to limit the feeding pressure of the herbivore. After many such cycles. the numbers of the herbivore populations are ultimately limited. and stability results.

Through the functioning of the genetic feedback mechanism. resistant factors in a given plant can be used to control a parasite which feeds on it. For example, on a susceptible plant genotype the animal population feeding heavily may be reproducing at a rate of two offspring per individual in the population. Under these conditions, the animal population would increase rapidly and would soon cause severe damage to the plant population by overfeeding. On the other hand, if resistant genes were concentrated in the plant population so that only resistant genotypes dominated, animal reproduction might be at a rate of one-half offspring per individual in the population. Then the animal population would decrease, and the damage to the plant population would be kept to a minimum.

An example of this type of change in a plant population took place in the Kansas wheat crop which was susceptible to the Hessian fly (*Phytophaga destructor*). As a result of the low resistance of the wheat to attack, Hessian fly populations increased, and the wheat crop suffered damage. With R. H. Painter's (23) development of a resistant variety of wheat, reproduction on the resistant wheat dropped to less than one per individual. and soon the fly population declined to a low level. Thus, by manipulation of the genotypes found in the wheat and not of the quantity of wheat, the fly population was controlled.

Although the interactions of wheat and Hessian flies can be considered to be man-made, we find evidence that under natural conditions biotic communities develop their own controls. In fact, populations in nature are usually regulated by several mechanisms that operate interdependently. These include not only genetic feedback but competition, parasitism, predation. and environmental heterogeneity.

This can be illustrated by a study of what might happen when a new animal species is introduced into a biotic community and becomes established on a plant. At first, the animal increases rapidly on its new plant host and reaches outbreak level. Under these conditions, competition for food among the animals is intense. In addition, the severe feeding pressure tends to eliminate many of the plants; this results in an altered distribution of the plants. With the plant hosts more sparsely distributed, the animal has increased difficulty in locating hosts, and some hosts have time to grow, reproduce, and maintain themselves.

Thus, at the early stages of interaction between animal and plant, competition and environmental heterogeneity, along with the pressure from parasites and predators, frequently limits the numbers of the animal and prevents the complete destruction of the host. If these factors are successful, then slower acting genetic change and evolution can take place.

Genetic change in the plant takes several generations because plant response to selective pressure exerted by the animal is slow. When a large animal population exerts severe feeding pressure on the plant population, large numbers of plants are destroyed. The first plants destroyed are primarily those most susceptible to the feeding pressure of the animal; the surviving plants generally carry one or more resistant genes. Under natural conditions, the evolution that occurs in the plant would rarely be caused by mutation but would be due to a recombining and concentrating of the genes already existing at low frequencies in the plant population. Resistance in the host is generally polygenic, and evolution proceeds slowly as genes are recombined in individuals and concentrated in the population. For example, there might be twenty loci (two genes per locus) in the host plants for some resistant character such as hardness. As susceptible genes at each locus are slowly replaced with resistant genes, the amount of resistance gradually increases.

When we look at the problem from a different angle, we find that the change and resistance in the plant can be measured as a response in the survival of the animal; that is, the number of eggs produced, the rate of development and growth, mortality, and longevity of the animal might all be influenced by increasing the concentration of resistant genes in the plant host. At a critical level of resistance in the host, the low birth rate and high death rate in the animal population would result in a significant decrease in numbers, and eventually the population would be sparse. Then with animal numbers rare in relation to those of the plant host, the animal population would only be removing "interest" (excess individuals or energy, or both) from the plant population, and relative equilibrium would exist between plant and animal. The animal would no longer be removing "capital" (those individuals or that energy, or both, needed for maintenance of the plant population). Evolution of this kind with a balance between supply and demand is possible with the genetic feedback mechanism.

Feeding pressure of herbivores, parasites, and predators on their plant or animal host may be limited by various protective mechanisms in their host, but there are examples of subtle genetic changes that significantly affect the survival of the animal that uses the host plant or host animal for food. For instance, when young pea aphids (*Acrythosiphum pisum*) were placed on a common crop variety of alfalfa (*Medicago sativa*), they produced a mean of 290 offspring in 10 days, whereas the same number of aphids for a similar period on a resistant alfalfa variety produced a mean of only 2 offspring (24). In another example, the mean rate of oviposition (eggs per generation) of the chinch bug (*Blissus leucopterus*) on a susceptible strain of sorghum (*Sorghum vulgare*) was about 100, whereas on a resistant strain the mean oviposition was less than one (25). In both, reproduction in the animals feeding on the resistant plant hosts decreased more than 99 per cent. This reduced reproduction obviously would have dramatic effects on the population dynamics of the feeding animal populations.

Resistance is effective in limiting animal numbers, and evidence suggests that it plays a dominant role in controlling populations in nature. If so, this would explain why population outbreaks occur frequently in newly introduced species. With little or no resistance, the new species increases rapidly on its susceptible food hosts. Until resistance in the plant host gradually increases, both outbreaks and intense fluctuations will occur. When relative stability is eventually reached and resistance is fully effective, animal numbers will be low. This is one reason why most animals are rare and especially rare relative to their food resource.

In addition, the relative stability and responsiveness of living systems are believed to account in part for the fact that most animals feed on living matter. The interaction between eating and eaten species and genetic feedback within the community form a complex but fully responsive system. Living systems, of course, respond to change and can evolve to provide a functional system whereby careful control of supply and demand can be achieved within the community as a whole.

The adaptation of supply by the plant and demand by the animal evolves and in time attains a state of relative balance. The plant host responds and evolves to its attacking animal only if the numbers of the animal are sufficient to exert some selective pressure on the host. This means that the trophic interactions between herbivore and plant, parasite and host, and predator and prey are important in determining the structure of the community. Based on this knowledge, Elton's statement that "the whole structure and activities of the community are dependent upon questions of food supply" takes on great significance in population control.

## PARASITE-HOST SYSTEMS

The validity of genetic feedback functioning as a regulatory mechanism in populations was investigated under controlled laboratory conditions (26). The premise of the first experiment was that the numbers of the feeding species would be controlled as genetic resistance evolved in the host population. The housefly (*Musca domestica*) was the host species, and a wasp (*Nasonia vitripennis*) was the parasite or feeding species. These two species were allowed to interact in the experimental unit for 1,004 days while host numbers were kept constant and parasite numbers were allowed to vary. The control unit was similar in design, except that hosts for the parasite population came from a population of houseflies that had not been exposed to the parasite. Hosts that survived exposure to the control parasites were destroyed to prevent the control host population from evolving. In both the control and experimental units, all parasites that emerged from their host types were saved and were returned to their respective population cages.

During the period of study, measurable evolution took place in both the host and parasite populations in the experimental unit. The experimental host population became more resistant to the parasite, as evidenced by a drop in the average reproduction of 135 to 39 progeny per experimental female parasite and a decrease in longevity from about 7 to 4 days. Concurrently, the parasite population evolved some avirulence toward its host. As the experiments progressed, selective pressure on the experimental host population declined, and density of the parasite population declined to about one half that of the control (about 3,700 for the control and 1,900 for the experimental). The amplitude of the fluctuations of experimental population (Figure 1) was significantly less than those experienced by the control.

The ecology and evolution of this same parasite and host were investigated in another experiment during which both parasite and host density were allowed to vary. A specially designed cage,

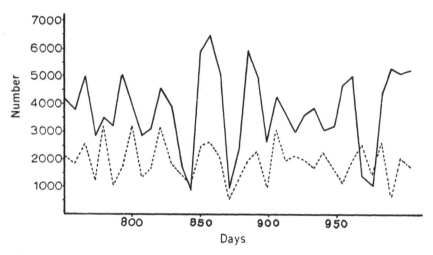

FIGURE 1: Population trends of parasite populations for the last 254 days of the 1,004-day period of two laboratory parasite-host systems. Solid line, control parasite; dashed line, experimental parasite.

consisting of thirty plastic cells joined together to make a multi-celled structure, provided space-time structure for normal parasite-host interactions. With this cage, the population characteristics exhibited by the control or newly associated parasite-host system were compared with those of the first experimental system in which some ecological balance had evolved.

During the 581-day period for the control system and 322-day period for the experimental system, parasite numbers averaged 118 per cell in the control system and only 32 in the experimental system. Host numbers averaged 172 per cell in the control and 462 in the experimental (Figure 2). Population fluctuations in the control system were severe, whereas in the experimental system they were dampened. The greater stability which the experimental system had already attained enabled it to make efficient use of its environmental resources and in this way increase its chances for survival.

One of the outstanding examples of the genetic feedback functioning in a natural population is the relationship of myxomatosis virus and European rabbits in Australia. After its introduction there in 1859, the European rabbit (Oryctolagus cuniculus) population increased to outbreak levels within the following 20 years (27). To reduce the density of the rabbit to a harmless level, the myxomatosis virus obtained from South American rabbits was introduced into the rabbit population. In essence, this action was analogous to introducing a new virus species into another community, for the myxomatosis virus and European rabbit had never been associated before. The virus spread rapidly in the rabbit population and immediately reached outbreak levels. During the first epidemic, myxomatosis was fatal to about 98 per cent of the rabbits; the second epidemic resulted in about 85 per cent mortality; and by the sixth epidemic, mortality was about 25 per cent (28). Today the virus is less effective than it had been but is still taking its toll of rabbits. Fenner summarized the situation by stating, "We could then envisage a climax association in which myxomatosis still caused moderately severe disease with an appreciable mortality, much

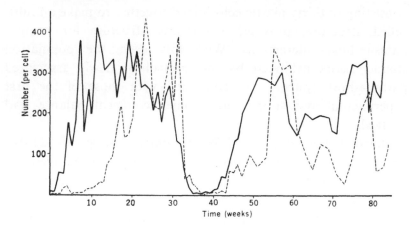

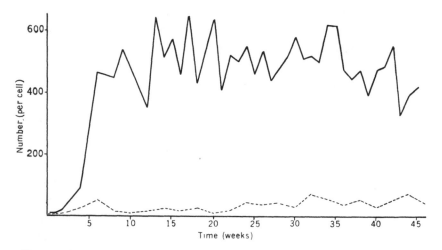

FIGURE 2: Population trends of control parasite-host system (top) and experimental system (bottom) that has evolved some stability or balance between the interacting species. Solid line, mean number of hosts; dashed line, mean number of parasites per cell in 30-cell population cages.

as smallpox does in human communities. The reproductive capacity of the rabbit is such that this sort of disease need not seriously interfere with its population size."

In this adjustment between virus and rabbit, attenuated genetic strains of virus evolved by mutation and tended to replace the virulent strains (29). In addition, passive immunity to myxomatosis is conferred to kittens born of immune does (30). Finally, a genetic change has occurred in the rabbit population, and this has provided intrinsic resistance to the myxomatosis virus (31). This clearly illustrates the alternate functioning of the feedback of density, selection, and genetic change which has in turn altered the density of both populations. There was some similarity between the virus-rabbit relationship, the laboratory wasp-fly relationship, and the type of evolution which took place. In the virus-rabbit association, most of the evolution occurred in the parasite, whereas in the wasp-fly association most of the evolution took place in the host.

Transmission of the myxomatosis virus depends upon mosquitoes (*Aedes* and *Anopheles*) that feed only on living animals (32). Rabbits infected with the virulent strain of virus live for a shorter period of time than those infected with the less virulent strain. Because rabbits infected with the less virulent strain live for longer periods of time, mosquitoes have access to that virus for longer periods of time. This gives the avirulent strain a competitive advantage over the virulent strain. In addition, in regions where the avirulent strain is located, rabbits are more abundant, and this allows more total virus to be present than in a comparable region infected with the virulent. Thus, the virus with the greatest rate of increase and density within the rabbit is not the virus selected for, but the virus with demands balanced against supply has survival value in the ecosystem.

Another example of how population regulation evolves from one dominant mechanism to another can be found in a comparison of the results of the initial interaction of the parasite-host study with the results of these populations after they had interacted for 1,004 days (26). Initially, parasite density in the experimental system averaged about 3,700. Although the density of the parasite population fluctuated, the mean reproduction of a parasite pair at the carrying capacity of the environment would have to be 2 or a pair with births equaling deaths. Because the experimental parasites produced about 135 progeny per female, 133 of these would have to die each generation to leave a single parasite pair surviving to replace the parent pair. Early in the experimental system, competition was primarily responsible for limiting parasite numbers and causing the death of 133 of the 135 offspring produced per female. The decline from 135 to 39 progeny per female of the experimental parasite meant that the loss of 96 progeny was due to changes brought about by genetic feedback. To maintain the population at this lowered reproductive rate, only 37 of the progeny could be lost to competition. Thus, competition in the beginning was the dominant control mechanism operating in the experimental system, but genetic feedback became dominant with time and through evolution.

## COMPETITION AND COEXISTENCE

When we consider how the genetic feedback mechanism functions, it seems logical to apply it to situations in which competing species might evolve to occupy the same niche. Competition here refers to species at the same time and place which share the same essential resource in short supply (2). Niche is defined as an animal's "place in the biotic environment, its relationship to food and enemies" in the community (17).

Competing species seeking the same plant, prey, or host can coexist if their numbers are controlled by genetic feedback. For example, let us assume that two aphid populations feed on sap from the same plant species. The two aphid species can coexist because the more abundant aphid species will eventually be controlled through the processes of genetic feedback. The amount of change that occurs in the characteristics of the plant for protection against the feeding pressure of the animal is dependent on density. Because more plants are selectively destroyed by the abundant aphid, the resistant polygenic factors effective against the abundant aphid would increase in the plant population. This means that the abundant aphid ultimately will be more limited by changes in the plant than the sparse aphid will. Thus, the numbers of both competing aphid populations are controlled by differential evolution of the plant relative to each population.

Results of field studies with two aphid species that attack alfalfa (33) suggest that two competitive animal species seeking the same food host can be differentially influenced by evolution in the plant.

Genetic feedback may also operate in yet another way to enable two species to coexist and utilize the same thing (food, space, and so on) in the ecosystem. In this case, let us assume that both species are fairly evenly balanced in their competitive ability and that species A is only slightly superior to species B. As the numbers of A are increasing, the numbers of B will be declining and becoming sparse. The abundant individuals of species A must contend principally with intraspecific competitive selection because there is a greater chance for individuals of this species to interact with their own kind. Haldane (34) pointed out that intraspecific competitive selection is frequently biologically disadvantageous for the species. At the same time, individuals of species B are contending primarily with interspecific competitive selection. Thus, under this selection species B would evolve and improve its ability to compete with its more abundant cohort species A. As species B improves as a competitor, its numbers increase, and finally B becomes the more abundant species. Then the dominant kind of competition (interspecific or intraspecific) affecting each species is reversed. After many such oscillations and with each oscillation decreasing in intensity, a state of relative stability should result.

This idea — that intraspecific selection on the dominant species and interspecific selection on the sparse species favors the sparse species — was tested successfully with the housefly and blowfly (*Phaenicia sericata*) in a multicelled cage (35). In another population system (surviving for 160 weeks or 80 fly generations), there was a persistent alteration of dominance of first the blowfly and then the housefly. A genetic check on the fly populations showed that the currently dominant species remained genetically static, while the sparse species or "underdog" evolved to become the better competitor and dominant species. Although there has been an oscillation in dominance, no damping of the fluctuation has been noted to date.

## CONCLUSION

The importance of the genetic feedback mechanism as a regulatory system in communities is substantiated by its wide application to such diverse interacting population systems as herbivore and plant, parasite and host, predator and prey, and interspecific competitor systems. The real significance of this mechanism for population regulation lies in the fact that it has its foundation in evolution. Population regulation by genetic feedback supports Emerson's (36) view that evolution in natural populations is toward homeostasis (balance) within populations, communities, and ecosystems.

Students of population ecology and especially of parasitology and epidemiology generally accept the fact that evolutionary trends in relationships of parasite and host are toward balance. The deductive basis for this generalization rests on the ecological principle that disharmony results in serious losses to both parasite and host. Large numbers of fatal infections in the host population eventually lead to host extinction which in turn brings about the extinction of the parasite. The success of any living population is measured by its relative abundance and distribution as well as its ability to survive in time.

Homeostasis, in herbivore-plant, parasite-host, and predator-prey species and among other community members in general, results in improved survival of the community system. The evolved balance in supply and demand achieved by the feeding species and its host establishes a sound economy for the community. This, of course, enables the community to make effective use of the resources available to it.

Increased species diversity in a community is due in part to community homeostasis. The genetic integration of interspecific competitors which makes possible the use of the same resource by competing species and enables them to occupy the same niche contributes to greater species diversity. The increased network of interactions within the community, resulting from a greater number of species present, further contributes to community homeostasis.

With more knowledge concerning the regulation of natural populations. man will be in a better position to control the pests on his food crops and the parasitic diseases of mankind. This will also help conserve the millions of living species which are vital for the functioning of the vast living system of which he is a part.

## REFERENCES

1. D. Lack, *The Natural Regulation of Animal Numbers* (Clarendon Press, Oxford, 1954).
2. A. MacFadyen, *Animal Ecology* (Pitman, London, 1957).
3. F. A. Urquhart, *Changes in the Fauna of Ontario* (Univ. of Toronto Press, Toronto, 1957).
4. L. L. Snyder, in *Changes in the Fauna of Ontario*, F. A. Urquhart, Ed. (Univ. of Toronto Press, Toronto, 1957), pp. 26–42.
5.———, *Ontario Birds* (Clarke, Erwin, Toronto, 1951).
6. O. Kalela, *Bird-Banding*, 20, 77 (1949).
7. P. P. Feeny, thesis, Oxford University (1966).
8. J. M. Kingsbury, *Poisonous Plants of United States and Canada* (Prentice-Hall, Englewood Cliffs, N.J., 1964).
9. T. Eisner, *Science*, 146, 1318 (1964).
10. ——— and J. Meinwald, *ibid.*, 153, 1341 (1966).
11. B. F. Coon, R. C. Miller, L. W. Aurant, *Pennsylvania Agricultural Experiment Station Report* (1948).
12. J. H. Pepper and E. Hastings, *Montana Agricultural Experiment Station Technical Bulletin*, 413 (1943).
13. H. G. Andrewartha and L. C. Birch, *The Distribution and Abundance of Animals* (Univ. of Chicago Press, Chicago, 1954).
14. H. S. Smith, *Econ. Entomol.*, 28, 873 (1935).
15. R. A. Fisher, A. S. Corbet, C. B. Williams, *J. Anim. Ecol.*, 12, 42 (1943).
16. E. R. Dunn, *Ecology*, 30, 39 (1949).
17. C. Elton, *Animal Ecology* (Sigwick and Jackson, London, 1927).
18. A. P. Jacot, *Ecology*, 17, 359 (1936).
19. C. Overgaard, *Natura Jutlandica*, 2, 1 (1949).
20. R. N. Chapman, *Animal Ecology* (McGraw-Hill, New York, 1931).
21. M. Demerec, *Biology of Drosophila* (Wiley, New York, 1950).
22. D. Pimentel, *Amer. Natur.*, 95, 65 (1961).
23. R. H. Painter, *Insect Resistance in Crop Plants* (Macmillan, New York, 1951).
24. R. G. Dahms and R. H. Painter, *J. Econ. Entomol.*, 33, 482 (1940).
25. R. G. Dahms, *J. Agr. Res.*, 76, 271 (1948).
26. D. Pimentel and R. Al-Hafidh, *Ann. Entomol. Soc. Amer.*, 56, 676 (1963).
27. D. G. Stead, *The Rabbit of Australia* (Winn, Sydney, Australia, 1935).
28. F. Fenner, in *The Genetics of Colonizing Species*, H. G. Baker and G. L. Stebbins, Eds. (Academic Press, New York, 1965), pp. 485–499.
29. H. V. Thompson, *Ann. Appl. Biol.*, 41, 358 (1954).
30. F. Fenner, *Cold Spring Harbor Symp. Quant. Biol.*, 18, 291 (1953).
31. I. D. Marshall, *J. Hyg.*, 56, 288 (1958).

32. M. F. Day, *J. Australian Inst. Agr. Sci.*, 21, 145 (1955).
33. R. H. Painter, *Proc. Int. Congr. Entomol. 12th*, 1964, 531 (1964).
34. J. B. S. Haldane, *The Causes of Evolution* (Longmans, Green, New York, 1932).
35. D. Pimentel, E. H. Feinberg, P. W. Wood, J. T. Hayes, *Amer. Natur.*, 99, 97 (1965).
36. A. E. Emerson, in *Principles of Animal Ecology*, W. C. Allee, A. E. Emerson, O. Park, T. Park, and K. P. Schmidt, Eds. (Saunders, Philadelphia, 1949), pp. 640–695.

# Index

Milton Keynes UK
Ingram Content Group UK Ltd.
UKHW040013071024
449327UK00011B/210